BestMasters

Mit „**BestMasters**" zeichnet Springer die besten Masterarbeiten aus, die an renommierten Hochschulen in Deutschland, Österreich und der Schweiz entstanden sind. Die mit Höchstnote ausgezeichneten Arbeiten wurden durch Gutachter zur Veröffentlichung empfohlen und behandeln aktuelle Themen aus unterschiedlichen Fachgebieten der Naturwissenschaften, Psychologie, Technik und Wirtschaftswissenschaften. Die Reihe wendet sich an Praktiker und Wissenschaftler gleichermaßen und soll insbesondere auch Nachwuchswissenschaftlern Orientierung geben.

Springer awards "**BestMasters**" to the best master's theses which have been completed at renowned Universities in Germany, Austria, and Switzerland. The studies received highest marks and were recommended for publication by supervisors. They address current issues from various fields of research in natural sciences, psychology, technology, and economics. The series addresses practitioners as well as scientists and, in particular, offers guidance for early stage researchers.

Weitere Bände in der Reihe http://www.springer.com/series/13198

Denise Drozd

Topographic Organization of the Pectine Neuropils in Scorpions

An Analysis of Chemosensory Afferents and the Projection Pattern in the Central Nervous System

 Springer Spektrum

Denise Drozd
Institute of Neurobiology
Ulm University
Ulm, Germany

ISSN 2625-3577   ISSN 2625-3615   (electronic)
BestMasters
ISBN 978-3-658-25154-3   ISBN 978-3-658-25155-0   (eBook)
https://doi.org/10.1007/978-3-658-25155-0

Library of Congress Control Number: 2018967683

Springer Spektrum

This Springer Spektrum imprint is published by the registered company Springer Fachmedien
Wiesbaden GmbH part of Springer Nature
The registered company address is: Abraham-Lincoln-Str. 46, 65189 Wiesbaden, Germany

Acknowledgement

I would like to thank Prof. Dr. Harald Wolf for giving me the opportunity to work on this intriguing topic. He inspired me through his interesting lectures and the discussions afterwards and encouraged me to work with arthropod nervous systems.

A big thank you to Dr. Torben Stemme, who supported and helped me a lot at the beginning of my work and encouraged me in tough times. He gave me so many tips and showed me tricks in the lab, which sped up the whole process. Furthermore, he always made time for me when there was something to discuss or just to drink a coffee together. Thank you so much!

Many thanks to my colleagues in the Institute of Neurobiology, especially PhD candidates Anja Dünnebeil, Janina Hladik and Dr. Sarah Pfeffer with whom I had a great time talking, discussing and laughing. I also like to thank the staff in the Institute for always helping and caring when I seemed lost with something.

This thesis wouldn't be possible without the care and support of my family and friends. They were always there for me and helped me in difficult times.

Thank you all for making this possible.

Contents

Table of Figures

Abstract

While most arthropods utilize their antennae to perceive chemosensory input, chelicerates evolved different organs for chemo-sensing. Scorpions (Koch, 1837) have pectines, which are specialized comb-like structures located on the ninth body segment and are used to probe the substrate for chemo- and mechanosensory information. They are subdivided into smaller units, the so-called pegs. Afferents of these sensory appendages project into a distinct neuropil of the central nervous system, located behind the fourth walking leg neuropils. However, detailed neuroanatomical data concerning sensory projections into the nervous system are still missing, but crucial for functional considerations. In this study, afferents of single pegs of *Mesobuthus eupeus* were analyzed by backfilling techniques, combined with immunohistological labeling of neuropilar regions. Staining of neuropil areas revealed the glomerular and lobular organization of the primary posterior pectine neuropil and a second homogenously structured anterior neuropil. The latter extended anteriorly near the ganglion midline up to the level of the second walking leg neuromeres. Posterior and anterior neuropils were shown to be bilateral structures, receiving input from the respective ipsilateral pectine. The large lobulus area consisted of large overlapping lobuli, was in juxtaposition to the posterior pectine neuropil in anterior direction and connected both pectine neuropils. Sensory projections of each peg entered the posterior pectine neuropil on the ventral side and innervated distinct parts of the primary neuropil: Afferents of distal pegs projected into lateral areas, sensory fibers of medial pegs innervated central areas of the neuropil and axons of proximal pegs projected into medial areas. After leaving the posterior pectine neuropil through a medially located tract, the axon bundles terminated in an anterior neuropil, with no distinct somatotopic distribution. The somatotopic organization of chemosensory afferents in the primary pectine neuropil suggests that the peg arrangement serves for differential perception of chemical gradients on the substrate, which might support orientation based on substrate-born signals.

1 Introduction

Arthropods are the most diverse and successful phylum and emerged presumably during the Cambrian explosion, approximately 540 Mya (Butterfield, 2003). Although varying in morphological features, behaviors, habitats and lifestyles, all hexapods, chelicerates, crustaceans and myriapods share a common ground pattern. They are characterized by jointed limbs, a segmented body and an exoskeleton, consisting of a chitin cuticle (Cutler, 1980; Schmidt-Nielsen, 1984; Rupert, 2004). The cuticle serves as a protection against dehydration, gives mechanical stability to the body and allows to interact with the environment (Wainwright, 1982; Dzik, 2007; Wolf, 2017). Unlike vertebrates, which have a more or less flexible cuticle and therefore developed different types of extero- and proprioceptors (Romer and Parsons, 1977), the sensory receptors of arthropods are located within the cuticle and have special adaptations to remain responsive for external stimuli, like mechanosensory campaniform sensilla (Keil and Steinbrecht, 1984; Hallberg and Hansson, 1999). The limitation in flexibility of the cuticle led to the development of various sense organs in arthropods, such as the antennae in hexapods (Schneider, 1964), in myriapods (Keil, 1976; Tichy and Barth, 1992) and antennules in crustaceans (Schmitt and Ache, 1979), the lyriform and tarsal organs in spiders (Anton and Tichy, 1994; Barth, 2004), the chemosensory malleoli in solpugids (Brownell and Farley, 1974) or the pectines in scorpions (Carthy, 1966; Brownell, 1988).

1.1 Chemo- and mechanoreception in arthropods

Chemo- and mechanoreception is of pivotal importance for arthropods (Dethier, 1947), a fact documented by the wide variety of often well elaborated antennal structures in hexapods, crustaceans, as well as myriapods (Hansson and Stensmyr, 2011) and equivalent chemo- and mechanosensitive appendages found in chelicerates (Carthy, 1966; Brownell and Farley, 1974; Brownell, 1988; Weygoldt, 2000).

© Springer Fachmedien Wiesbaden GmbH, part of Springer Nature 2019
D. Drozd, *Topographic Organization of the Pectine Neuropils in Scorpions*,
BestMasters, https://doi.org/10.1007/978-3-658-25155-0_1

1.1.1   Chemosensitive structures in arthropods

The chemosensory systems are usually divided in olfaction and gustation, although both senses underlie similar concepts; in both cases chemicals have to access sensory dendrites by pores in the cuticle and further by transport proteins in the receptor lymph (Zacharuk, 1980; Steinbrecht, 1987; Tichy and Barth, 1992). Differences occur in the directness of chemical contact of gustatory and chemosensitive sensilla. Whereas gustatory sensilla usually need direct contact with the substrate, olfactory sensilla in the antennae are specialized for chemical plumbs, which are often lower in density. To ensure olfactory sensitivity, the external surface of the sense organ has to be enlarged by branching or by elongation. Such branching of the antennae can be observed e.g. in the silk moth *Bombyx mori* where males display great enlarged und branched antennae, to follow female pheromones in a zigzag flight (Kanzaki and Ikeda, 1994). Gustatory sensilla are often found on the mouthparts and the tarsi of arthropods (Foelix and Schabronath, 1983; Edgecombe and Murdock, 1992; Stocker, 1994; de Brito Sanchez et al., 2014), to probe the substrate for nutrients or for reproduction relevant features. Chemosensory sensilla in arthropods show a wide variety of different structures (hairs, pegs, loops or plates) (Schneider and Steinbrecht, 1968) with numerous taxon specific adaptations. However, they share the same basic features, such as the pore on the tip of the sensilla. This pore contains the sensory dendrites and allows contact with the environment by mediating contact of chemicals with the receptors and conveying the information to the central nervous system (CNS) (Slifer, 1970). Chemosensitive sensilla and other arthropod sensilla are built according to a common 'bauplan' and are probably homologous to mechanosensitive scolopidia (Schmidt, 1969). Arthropod sensilla are cuticular sensory organs and are composed of three basic structures: sensory cells, enveloping cells, and a cuticular structure (Hallberg and Hansson, 1999). A cilliar region, bearing a transformed cilium, possesses a dendrite with the bipolar cells and axons to the CNS. This cilium may exhibit different specializations depending on its function. The sensory cell is enveloped by the thecogen cell, which forms the dendritic sheath, the trichogen cell, which forms the hair and the outermost tormogen cell, which secretes the cuticle (Keil and Steinbrecht, 1984; Hallberg and Hansson, 1999).

1.1.2   Mechanosensitive structures in arthropods

Scolopidia are the most basic mechanosensitive structures, which consist of a scolopale cap, a scolopale cell structure, which envelopes the dendritic cilium and a bipolar sensory nerve cell. There are different structures and forms of scolopidia in arthropods. Differences concern the number of sensory cells and the connection of the dendritic outer segments to the scolopidial tip. The connections can be direct (amphinematic) or indirect (mononematic) (Hallberg and Hansson, 1999).

1.2 Processing of sensory information in the arthropod nervous system

The nervous system of arthropods is usually divided into the protocerebrum, deutocerebrum, tritocerebrum, the subesophageal ganglion and the segmented ganglia. In chelicerates the segmental ganglia of the prosoma are fused to a ganglion mass, the synganglion, which came with the fusion of the first head and thoracic segments (Hanström, 1923; Babu, 1965; Babu and Barth, 1984). The processing of chemo- and mechanosensory information from the respective antennae in Mandibulata (Myriapoda + Crustacea + Hexapoda) is accomplished by the deutocerebrum, which is associated with the first pair of antennae. This region involves a paired olfactory lobe and a paired posterior neuropil (Schachtner, Schmidt and Homberg, 2005; Sombke, Rosenberg and Hilken, 2011). The posterior neuropils process mechanosensory input and have different names within the groups of mandibulate arthropods. While in malacostracan crustaceans and Remipedia the posterior neuropils are cone-shaped and named lateral antennular neuropils and median antennular neuropils (Schmidt and Ache, 1992; Sandeman et. Al. 1992; Fanenbruck and Harzsch 2005; Harzsch and Hansson, 2008), the neuropils in Chilopoda are elongated, palisade-shaped neuropils and named corpus lamellosum (Hörberg, 1931; Fahlander, 1938; Sombke, Harzsch and Hansson, 2011; Sombke, Rosenberg and Hilken, 2011). In Hexapoda, mechanosensory neuropils are called the antennal mechanosensory and motor control center (Hanström, 1928; Homberg, Christensen and Hildebrand, 1989) and have also an elongated shape.

The antennal lobe of these arthropod groups processes the chemosensory input of the antennae and comprises three central neuron classes, besides the primary afferents: local interneurons, projection neurons and centrifugal neurons. Arborizations of the local interneurons are limited to the antennal lobe, whereas arborizations of the projection neurons project from the antennal lobe to higher order brain regions, such as the mushroom bodies in Hexapoda and Chilopoda (Homberg, Christensen and Hildebrand, 1989; Strausfeld et al., 1998) and the hemiellipsoid bodies in Crustacea (Wolff et al., 2012) in the protocerebrum. The arborizations of the centrifugal neurons are located outside of the antennal lobe and innervate also regions of the protocerebrum. The subunit of the antennal lobe are the olfactory glomeruli, which are dense neuropils (Homberg, Christensen and Hildebrand, 1989). Axons of olfactory sensory neurons terminate in these olfactory glomeruli and communicate with olfactory interneurons. Chemosensory input is first integrated within the olfactory glomeruli and later conveyed to higher brain regions in the protocerebrum via olfactory projection neurons (Schachtner, Schmidt and Homberg, 2005; Sombke, Rosenberg and Hilken, 2011). In hexapods, the olfactory glomeruli represent a chemotopic map, where one glomerulus exists for each receptor type (Galizia and Menzel, 2000, 2001; Eishten, 2002; Ignell and Hansson, 2005; Hansson and Stensmyr, 2011).

1.3   Sensory structures in arachnids

Arachnids are a subphylum of the chelicerates and are distinguishable from other arthropods by their eight legs, their first pair of appendages, called chelicerae, their pedipalps, which in some clades can be raptorial (Barth, 1985) and the tagmatization into the prosoma and the opisthosoma (Ruppert et al., 2007). Only scorpions have an elongated body form and the opisthosoma is divided into meso- and metasoma (Hjelle, 1990). Unlike other arthropods, arachnids do not have antennal structures, but evolved chelicerae and pedipalps and therefore carry most of the sense organs on other appendages (Foelix, 1985). The chelicerae are homologous to the antennae of the deutocerebrum in hexapods and the pedipalps to the second pair of antennae in crustaceans and bear most of the chemo- and mechanoreceptors. This is due to the fact, that these appendages are adapted for feeding, probing, locomotion and reproduction (Bristowe, 1958). In several arachnid groups the pedipalps evolved into specialized appendages, with different functions and therefore, other more posterior located appendages adopted chemo- and mechanosensory function (Wolf, 2017). This is true for e.g. scorpions, which use their pincer-like pedipalps for prey capturing and pre-intestinal digestion. Scorpions evolved comb-like pectines on the ventral body side, posterior to the walking legs, for chemo-and mechanosensory perception (Foelix, 1983; Brownell, 1988; Gaffin and Brownell, 1992, 1997; Wolf, 2008, 2017).

1.4   Mechano- and chemoreceptors in arachnids

The arachnid hair sensillum is, besides the thermo- and hygrorecptors on the tarsi (Anton and Tichy, 1994), the basic receptor structure and can be distinguished by the mode of innervation into two categories: mechanoreceptors, which have dendrites ending at the base of the hair and chemoreceptors, which have dendrites entering the hair shaft and interact with the external world through pores in the hair wall (Foelix, 1985). Mechanoreceptors can be further divided into tactile hairs, trichobothria, joint receptors and slit sensilla (lyriform sensilla), which play an important role in perception of vibrations and orientation (Pringle, 1955; McIver, 1975).

The chemoreceptors of arachnids are the contact chemoreceptors, pore hairs and the tarsal organs (Foelix, 1970; Foelix and Chu-Wang, 1973; Egan, 1976; Dumpert, 1978; Foelix, 1985), which are innervated by multiple dendrites and are found mostly on the tarsi and on the chelicerae (Foelix, 1970). They are used to probe the substrate for chemical information or to analyze food items (Foelix, 1985).

1.4.1 Sensory appendages in scorpions - the pectines

Scorpions are capable of perceiving and filtering different qualities of external information with specialized structures on their bodies, the pectines (Babu, 1965). They are located ventrolaterally on the ninth body segment, behind the walking legs (Wolf, 2008) and are bilateral comb-like appendages, which extend the body laterally and consist of a flexible spine and varying number of pectinal teeth (Brownell,1988; Gaffin and Brownell, 1997; Melville, 2000). This number varies between different species and can vary between sexes of the same species (Wolf, 2008). Due to the formation of pectinal teeth or pegs, which are directed to the ground, the ventral surface area for sensilla presentation increases and a higher chemosensory and mechanosensory sensitivity can be achieved. Pectines are already formed during embryonal development as limb-buds and detach later in the ontogeny from the body, becoming movable at the base. During the ontogeny, the number of peg sensilla on the pectines increases (Farley, 2001) and can reach up to 100,000 sensilla in adulthood (Wolf, 2008). The ability to move the pectines, similar to the walking legs, is due to motor neuron innervation and three joints in the pectine spine (Wolf and Harzsch, 2002a, b; Wolf, 2008, 2017), which proposes the idea of ancestral swimming and balancing in early Silurian representatives (Dunlop and Webster, 1999; Farley, 2001; Dunlop and Braddy, 2001). The motor neuron supply suggests a similarity to the walking legs, with three soma groups: an anterior soma group, a posterior ipsilateral soma group and a contralateral inhibitory motor neuron somata group (Wolf and Harzsch, 2002b; Wolf, 2008). The movement of the pectines and their mechanosensitive character is used by scorpions for adjusting the posture, presumably mate-trail behavior (Gaffin and Brownell, 1992, 2001; Tallarovic, Melville and Brownell, 2000; Melville, Tallarovic and Brownell, 2003), deposition of spermatophore

(Alexander, 1957, 1959), locating prey (Krapf, 1986), perceiving and distinguishing substrate texture (Abushama, 1966; Carty, 1966) and detection of obstacles (Boyden, 1978).

The pectines bear different sensory structures on their surface. Short hair sensilla and peg sensilla on the pectines are presumably both chemo- and mechanosensitive, whereby the chemosensory character dominates in peg sensilla (Gaffin and Brownell, 1992, 1997a, b; Gaffin, 2002, 2010). Similar to the rest of the scorpion's body, the pectines have interspersed trichobothria (Babu, Sreenivasulu and Sekhar, 1993; Kladt, Wolf and Heinzel, 2007) and are densely covered with peg sensilla (Wolf, 2008), which are similar to arthropod chemo- and mechanosensitive sensilla (Gaffin, 1994; Hallberg and Hansson, 1999). The shaft of the peg sensillum is about $2 - 10$ µm above the cuticle and 5 µm in diameter. Molecules can enter the peg sensillum through a slit opening at the tip, which connects to the lumen through a lymph section (Ivanov and Balashov, 1979; Foelix and Müller-Vorholt, 1983; Gaffin, 1994). Unlike other arthropod sensilla, each bipolar sensory neuron (receptor cells) has one sensory dendrite (Wolf, 2017). Near the base of the peg sensillum, a shorter dendrite terminates, which suggests the transduction of mechanosensory information to the CNS (Foelix and Müller-Vorholt, 1983). In the bilateral pectine nerve all sensory afferents from the sensilla are bundled and enter the posterior pectine neuropil ventrally from the posterior direction.

1.4.2   The nervous system of scorpions and the pectine neuropil

The nervous system of scorpions consists of three brain parts, a fused subesophageal ganglion and the unfused ganglia of the metasoma, which control the tail movement and perceive the sensory input from trichobothria located on the cuticle (Hanström, 1923; Babu, 1965; Gaffin and Brownell, 1997). The brain is divided into the protocerebrum, which forms the anterior part of the neuroaxis, the deutocerebrum, which is adjacent posteriorly to the protocerebrum and the large tritocerebrum, forming the bulging anterior ventral end. The optic neuropils, the mushroom bodies and the arcuate body are part of the protocerebrum. In Mandibulata, the bilateral mushroom bodies are suggested to integrate sensory modalities from the antennal lobe and from the optic neuropils, but data verifying if chemosensory information is integrated in the scorpion mushroom bodies is lacking (Strausfeld et al., 1998). The deutocerebrum is associated with the chelicerates and is considered to be homologous to the tripartite mandibulate deutocerebrum (Scholtz and Edgecombe, 2006). Ventrally to the deutocerebrum the tritocerebrum is situated, through which the esophagus passes through. The pedipalp neuromeres, four prosoma walking leg neuromeres, as well as the four metasoma neuromeres of the genital and the pectine segments, are situated within the fused subesophageal ganglion. The size of the respective neuromeres correlates with the number of sensory and motor control neurons, which is why the pedipalp neuromeres are considerably bigger than the walking leg neuromeres. Despite the fact, that the pectines are much smaller as the walking legs, the pectine neuropils appear to be relatively large (see Fig. 1), because the pectine neuropils receive a great amount of chemo- and mechanosensory input from the pectines (Wolf, 2008). The afferents of the sensory structures project via the respective pectine nerve into the dense glomerular pectine neuropils from the ventral side. A small tract of sensory axons diverges from the main tract, extends anteriorly (Wolf, 2017).

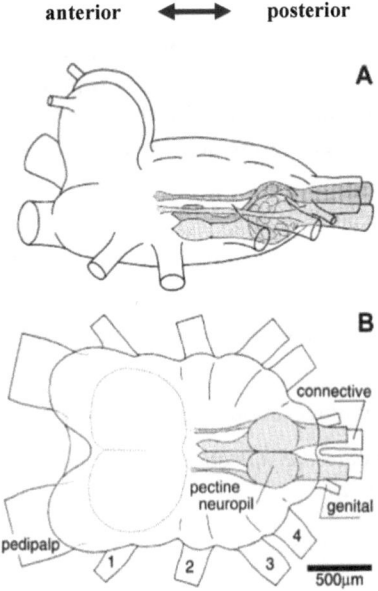

Figure 1: The fused nervous system of the scorpion (without the metasomal ganglia) and the location of the pectine neuropil. **(A)** Dorsolateral view of the synganglion. **(B)** Ventral view of the scorpions synganglion. Main nerve roots, which supply the pedipalps and the four walking legs are implied. Pectine neuropil is located at the posterior end of the fused prosomal ganglion mass. Drawing altered with permission of Prof. Dr. Harald Wolf (Wolf, 2008).

The size of the pectine nerve, as well as the pectine neuropil itself, depends on species and sex (Foelix and Müller-Vorholt, 1983; Wolf, 2008). Arborizations of afferents enter and envelop the pectine neuropil. Unlike the glomeruli in the antennal lobe, the glomeruli of the pectine neuropil seem to be more flattened and layered (Wolf, 2008). The size and number of these 'glomerular' compartments can vary among the species, but results of previous studies showed up to 140 'glomerular' structures in 19 layers in *Buthus occitanus* (Drozd, 2014; Hüll, 2014). Interneuron tracts extend longitudinally from the posterior pectine neuropil anteriorly to the level of the fourth and third walking leg neuromeres and terminate in a smooth anterior pectine neuropil, which do not seem to bear any compartmentalization (Wolf, 2008).

1.5   Goal of this thesis

This thesis illuminates the morphology of the pectine neuropil network and its relay stations within the nervous system of the scorpion *Mesobuthus eupeus* by immunohistochemistry and microscopy. Furthermore, the sensory innervation from single different located pegs were analyzed by retrograde tracing with Neurobiotin™.

Following questions shall be answered:

- What are the morphological features of the pectine neuropil and adjacent structures?

- Where do differently located pegs project within the posterior pectine neuropil?

- Is there a topographic innervation of the posterior pectine neuropil?

- Is it possible to compare the morphology and the projection pattern of the pegs with different taxa?

2 Material and methods

2.1 Animals – *Mesobuthus eupeus*

The scorpions of the species *Mesobuthus eupeus* (Koch, 1839), which were used in this study, were obtained from the online shop „the pet factory"(www.thepet-factory.de, Huelsede, Germany). The adult animals were kept together in tanks of the size of 33.5 x 20 x 18 cm under natural light conditions and temperatures between 22°C and 25°C. The tanks were filled with approx. 5 cm of sand as substrate and stones and bricks served as hiding places. Small shallow water dishes were provided and feeding occurred twice a week with small crickets (L3) (*Gryllus bimaculatus*) from the online store „born to be eaten" (www.futter-tiere24.de; b.t.b.e Insektenzucht GmbH, Schnuerpflingen, Germany).

2.2 Backfill of afferents fibers

Borosilicate glass capillaries (Harvard Apparatus LTD., USA) of the type GC100TF – 10 (1.0 mm O.D. x 0.78 mm I.D) were pulled with a pipette puller (Sutter Instrument CO., Model P-97, USA). Subsequently, the tip of the glass capillary was broken off to fit the pectinal tooth and filled with 5 % Neurobio-tin^{TM} Tracer (Vector Laboratories, Canada), dissolved in double-distilled water (bidest. H_2O).

For the backfill analysis, the animals were anaesthetized with 99.7 Vol. % CO_2 and subsequently cooled down at 4 °C in the refrigerator for 30 minutes. All following preparations were done with use of a binocular for better visualization. The cooled animals were fixed and immobilized dorsally onto cork plates with plasticine, so the ventral appendages were accessible for following preparations (Fig. 2). The pectines were also immobilized with wax and cleaned with demin-eralized water (demin. H_2O). One pectinal tooth on each pectine of an individual was cut apically with sharp small scissors and rinsed with demin. H_2O for 5 minutes. For the experiments, a second to last distal, one medial or the second most proximal pectinal tooth was cut and backfilled (see Fig. 2). Afterwards, the water was removed with a small piece of tissue and the Neurobiotin^{TM} filled glass

© Springer Fachmedien Wiesbaden GmbH, part of Springer Nature 2019
D. Drozd, *Topographic Organization of the Pectine Neuropils in Scorpions*,
BestMasters, https://doi.org/10.1007/978-3-658-25155-0_2

capillary was put onto the previously cut pectinal tooth. To prevent the setup from drying, petrolatum (Vaseline® Original, Germany), was applied on top of the pectine as well as the back of the capillary. The treated animals were stored in moist chambers for 3 days in the refrigerator at 4 °C (see Fig. 2).

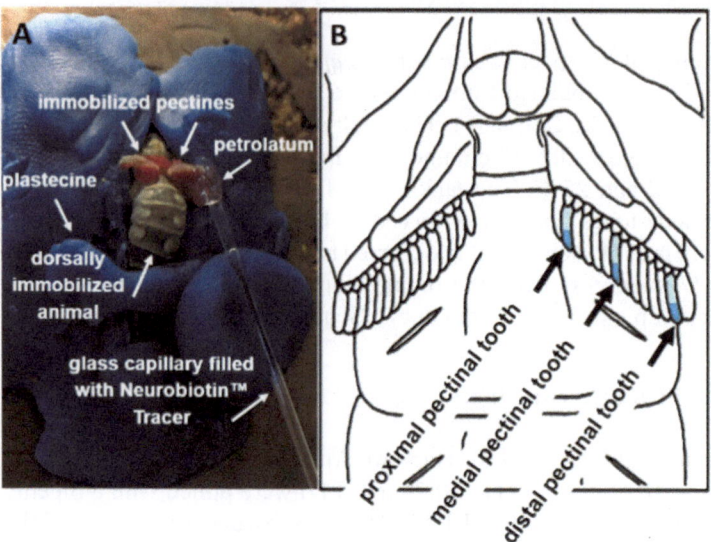

Figure 2: Dorsally immobilized and secured scorpion in the experimental setup and ventral view of the pectines. **(A)** Animals, as well as the pectines were immobilized with plasticine and wax. Glass capillary filled with Neurobiotin™ tracer was placed onto the previously cut pectinal tooth. **(B)** Ventral view of the scorpion. Respective pectinal teeth used for retrograde tracing (light blue) and incisions for inserting micropipette (dark blue). Right drawing was used and altered with the permission of Prof. Dr. Harald Wolf (Wolf, 2008).

2.3 Dissection, embedding and sectioning

The dissection of the animals was performed with the help of a binocular. After the animals were taken from the moist chamber, the glass capillary, plasticine and wax were removed. Cuts were made along the lateral and anterior side of the body, just above the appendages (walking legs and mouth parts). Furthermore, the animal was cut frontally at the level of the first mesosomal segment. To remove the ventral plate of the animal and to liberate the synganglion with the pectine neuropil, the nerves of the appendages and the connectives between the

CNS and the first mesosomal ganglion were cut. The ventral plate with the attached CNS was carefully removed with the help of two forceps and transferred into a shallow glass bowl filled with phosphate buffered saline (PBS 0.1 M, pH 7.4). Afterwards, the synganglion was separated from the surrounding tissue and the ventral plate and rinsed with fresh PBS. To fixate the nervous tissue, 4 % paraformaldehyde (PFA) in PBS was filled in the bowl and incubated for 2 hours at room temperature (RT), followed by three washing steps with PBS of 15 minutes each. In the meanwhile, 7 % low-melt agarose (Carl Roth GmbH + Co. KG, Karlsruhe, Germany) was heated to approx. 40°C. After washing, the synganglia were transferred to small black scale pans, carefully dried with filter paper, and coated with Poly-d-Lysine (1 mg/ml in H_2O, Specialty Media, Merck KGaA, Darmstadt, Germany). The coated synganglia were placed into small plastic bowls and embedded with the prepared agarose. After the agarose was set and dry, the blocks were trimmed with a razor blade and sectioned by a Leica VT1000 S vibratome (Leica Biosystems Nussloch GmbH, Nussloch, Germany) into 75 μm thick horizontal sections. Slices were stored in a 12 well cell culture plate (Costar®, Corning Incorporated, USA) filled with ice cold PBS.

2.4   Immunohistochemistry and mounting

The embedded tissue was permeabilized with 0.3 % Saponin (Fluka Bio-Chemika, USA) in PBS containing 0.3 %Triton X-100 (Sigma-Aldrich, Germany) (PBS-T 0.3 %) for 1 hour at RT. Subsequently, the tissue was washed in PBS-T 0.3 % for 30 min. and blocked for 3 hours at RT with a blocking solution, consisting of 5 % normal goat serum (Vector Laboratories, Ca) in PBS-T 0.3 %. The first antibody anti-SYNORF1 (Developmental Studies Hybridoma Bank, University of Iowa, USA) (Klagges, et al., 1996) was diluted 1:30 in blocking solution and applied onto the sections for 3 hours at 4 °C. Anti-SYNORF1 is an antibody which binds synapsins, a protein family involved in the exocytosis process in vesicle membranes of synapses. After washing the sections 3 times for 15 minutes each with PBS-T 0.3 %, the second antibody goat-anti-mouse Alexa Fluor® 488 (Thermo Fisher Scientific Life Technologies GmbH, Darmstadt, Germany) was diluted 1:250 in blocking solution and applied onto the sections. Additionally, Cy3™-Streptavidin (1:250; Jackson ImmunoResearch Laboratories, Inc. USA) and 4',6-Diamidin-2-phenylindol (1:100; DAPI, Sigma-Aldrich Co. LLC., Hamburg, Germany) was added to the blocking solution and applied onto the tissue. DAPI is used to label cell nuclei, due to its ability to intercalate with the DNA. The tissue was then incubated over night at 4 °C. Three washing steps with PBS-T 0.3 % and a final step with PBS for 15 minutes each followed. The sections were mounted onto microscope slides and covered with Mowiol® 4-88 (Merck KGaA, Darmstadt, Germany) as coverslipping reagents.

2.5   Microscopy and image analysis

The fluorescence microscopy was performed with the Leica DM5500 B (Leica Biosystems Nussloch GmbH, Nussloch, Germany) and the software LAS AF. The photos were recorded with 10x, 20x and 40x lenses and scanned in steps of 1 µm, 0,75 µm and 0,5 µm, respectively. To examine the fluorescence recordings, the picture stacks were loaded into ImageJ 1.5 and edited. This editing included, changing brightness, contrast and color of all three channels to highlight the findings. Some pictures were reconstructed with the help of Gimp 2.8. After editing all relevant pictures, panels were built with the plugin FigureJ in ImageJ.

3 Results

Thirty-five scorpions were used for this neuroanatomical analysis, but only thirteen experimental setups showed successful results. Four of these thirteen animals were males, nine were females. Pegs of all scorpions were counted. Male scorpions had a mean number of 24.23 ± 0.058 (n=17) pectinal tooth per pectine, females however had 19.30 ± 0.583 (n=18) pectinal tooth.

The nomenclature of single structures was adapted from previous studies of J. Melville (Melville, 2000) and H. Wolf (Wolf, 2008), based on position within the neuropil and similarities in structure and appearance.

3.1 Establishment of methods

The scorpion's nervous system consisted of a supraesophageal ganglion, which was located dorsally in the anterior front of the animal, a subesophageal ganglion mass, consisting of six pairs of neuromeres, including the pectine neuropils and in posterior direction extending metasomal ganglia. The neuromeres of the synganglion were associated with the pectines, the walking legs and the pedipalps (Babu, 1965). The outer rind, as well as the ganglion midline of the neuromeres contained cell somata. The posterior pectine neuropil (PPN) appeared ovoid in shape and was situated ventrally in the posterior part of the scorpion's synganglion, after the fourth walking leg neuromeres (see Fig. 3). The pectine neuropil extended the scorpions ganglionic mass caudally and connected the metasomal ganglia by connectives.

The results of the first Neurobiotin™ backfill experiments were gathered from two individual animals (T7, T8) and showed a general image of the projection areas of pectinal teeth in the posterior pectine neuropil (See Fig. 3). It was possible to observe the successful backfill of sensory afferents of the proximal peg and distal peg in the PPN of the animal. Because no structural marker was used, the PPN was only visible by means of the backfill of the pectinal teeth. Labeled fibers of the proximal peg projected into medial areas near ganglion midline, whereas afferents of distal peg projected into lateral regions of the pectine neuropil (Fig. 3). The afferents of the pectinal teeth entered the PPN through

© Springer Fachmedien Wiesbaden GmbH, part of Springer Nature 2019
D. Drozd, *Topographic Organization of the Pectine Neuropils in Scorpions*,
BestMasters, https://doi.org/10.1007/978-3-658-25155-0_3

the pectine nerve ventrally from the posterior part of the synganglion and propagated dorsally in anterior direction. In ventral sections, the projection area of proximal pegs appeared to be vague and in close vicinity to the ganglion midline (see. Fig. 3A) and in dorsal sections it appeared to be more distinct and spherical in shape.

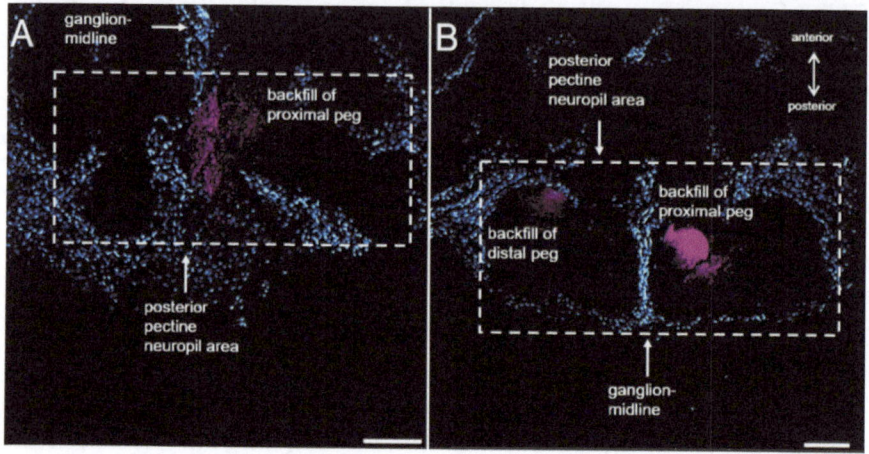

Figure 3: Neurobiotin™ backfill of proximal peg of the right pectine and distal peg of the left pectine of animal T7. Horizontal view. Ventral to dorsal sections. Backfill (BF) is shown in magenta, nuclei staining (DAPI) in cyan. Posterior pectine neuropil areas are indicated by white boxes in **(A)** and **(B)**. White arrows show the ganglion midline and the BF. **(A)** BF of afferents of the proximal peg projected near the ganglion midline of the right posterior pectine neuropil. **(B)** BF of the left distal peg projected into lateral areas of the pectine neuropil. Labeled afferents of the proximal peg in the right pectine projects near the ganglion midline. Scale bars: 100 μm.

The retrograde BF of a pectinal tooth exhibited an ovoid projection area near the left lateral margin of the pectine neuropil. The pectine neuropil area was situated below the fourth walking leg neuromeres (See Fig. 4). Sensory fibers of the distal peg innervated regions near the left lateral margin of the left pectine neuropil in anterior direction. This area was interrupted horizontally (shown by dashed line in Fig. 4A') and displayed a cap-like region and a smaller disc-like region below the division. Both regions were not visibly connected through tracts. Lateral located tracts proceeded along the anterior-lateral margin of the pectine neuropil in close vicinity to the somata cortex, and extended anteriorly in direction of the ganglion midline (see. Fig. 4).

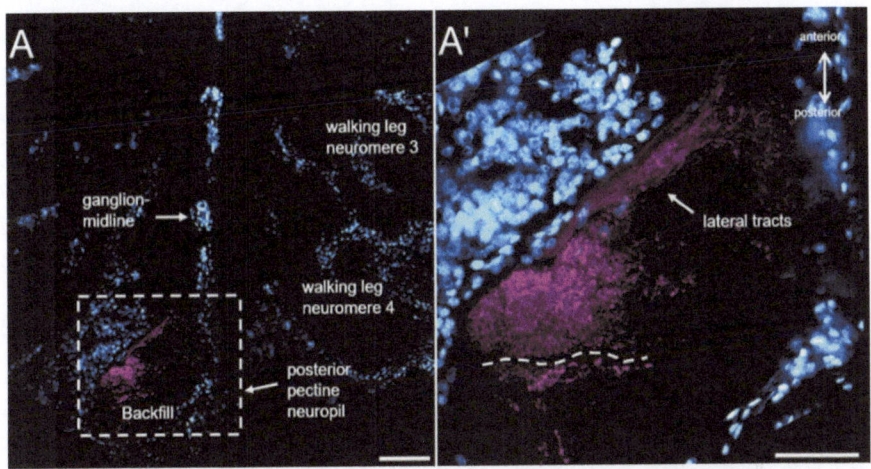

Figure 4: Neurobiotin™ backfill of distal pectinal tooth of the left pectine of animal T8. Horizontal view. BF is shown in magenta, counter staining of somata (DAPI) in cyan. Posterior pectine neuropil areas are indicated by white box in **(A)**. White arrows show the ganglion midline and the BF. **(A)** Posterior pectine neuropil situated behind the 4th walking leg neuromere. BF of afferents of the distal peg projected into lateral areas of the left posterior pectine neuropil. Scale bar: 100µm. **(A')** Inset of A. BF of the left distal peg projected into lateral areas of the pectine neuropil. White arrow indicates lateral tracts. Dashed line shows structural interruption. Scale bars: 50 µm.

3.2   Structural analysis of the posterior pectine neuropil and associated tracts

Sections were treated with anti-SYNORF1 antibody to identify synaptic tissue within the synganglion of the scorpion. Consequently, neuropilar regions were visible for structural analysis. To distinguish the respective structures within the pectine neuropil system, black appearing boundaries surrounding specific structures were used. This allowed to differentiate between lobular, lamellar and glomerular regions of the pectine neuropil and regions associated with the posterior pectine neuropil, like the ascending longitudinal tracts (see Fig. 5).

The immunofluorescence labeling revealed the clubbed PPN and its subdivisions, which consisted of the anterior located large lobulus area (LLA), a cap-like area and a disc-like area. The pectine neuropil showed lobular and lamellar structures; two visible lobules in the disc-like area, four lateral lobules, located each on the respective neuropil side and three layers of continuous glomeruli. The pectine neuropils were 233.08 µm ±: 8.379 in width and 226.57 µm ± 10.408

(n=9) in length. Longitudinal tracts were situated near the ganglion midline and had a length of 517.645 µm ± 6.461 (n=4). These tracts ended at the level of the third walking leg neuromere.

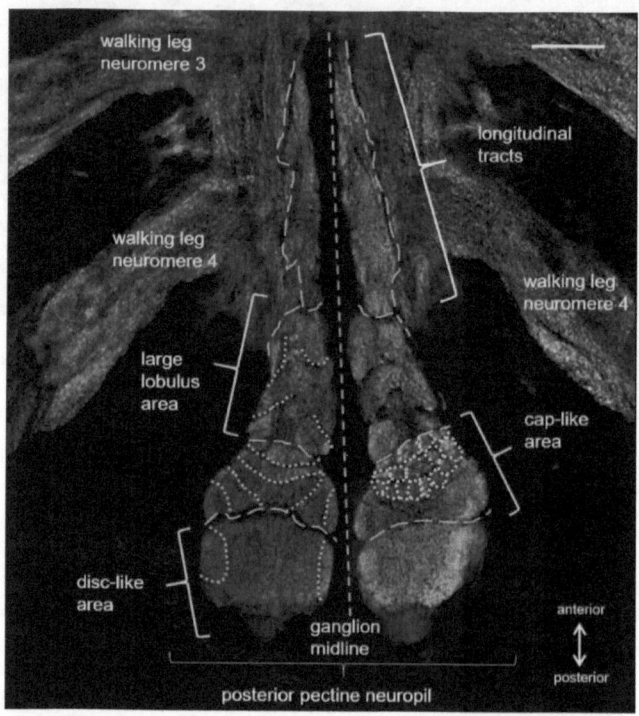

Figure 5: Anti-SYNORF1 labeling of neuropilar areas in the CNS of *M. eupeus* T31. Horizontal view. Synaptic tissue in grey. Fragmentation of the posterior pectine neuropil into four distinctive areas (disc-like area, cap-like area, large lobulus area) and anteriorly proceeding longitudinal tracts (white dashed lines). Lobular and lamellar structures are indicated by blue dotted lines in the left pectine neuropil and glomerular structures by yellow dotted circles. Vertical white dashed line indicates ganglion midline. Scale bar: 100 µm.

Closer inspection of the labeled pectine neuropil of *M. eupeus* T20 revealed additional features and structures, which are shown in Figure 6. When viewing the PPN of one body hemisphere in horizontal sections, the internal architecture appeared symmetrical (see. Fig. 6D).

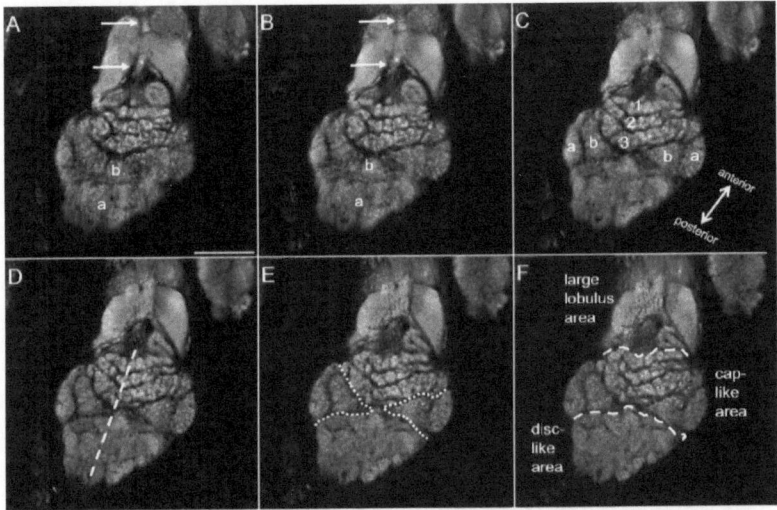

Figure 6: Close-up series of the internal structure of the PPN of *M. eupeus* T20 revealed by anti-Synapsin immunolabeling. Horizontal view. **(A-F)** ventral to dorsal sections. Sections taken at 6 µm intervals. Synaptic tissue in grey. **(A)** The lobules of the large lobulus area were attached to longitudinal proceeding axonal bundles (indicated by two arrows). **(B)** **(C)** Three rows of layers with glomerular structures, indicated by numbers and two lateral lobular structures indicated by a and b respectively could be distinguished. These lobules are connected ventrally in the disc-like area, shown in **(A)** and **(B)**. **(D)** Symmetrical superstructure of the PPN labeled by white dashed line. **(E)** Symmetrical lobular structures of the cap-like area were located laterally of the PPN. **(F)** The large lobulus area was located anteriorly of the PPN indicated by white dashed line. The cap-like area and the disc-like area were situated posteriorly of the large lobulus area. Scale bar: 100 µm.

The cap-like area and the disc-like area displayed together an ovoid shape and each subdivision showed a specific internal structure. Whereas the structure of cap-like area consisted of three horizontal glomerular layers, the disc-like area showed a palisade-shaped lobular fragmentation. The layers of the cap-like area resembled a bell-shaped curve and were distinguishable from the surrounding tissue, by their glomerular structure. The cap-like area exhibited symmetrical lateral lobular areas, with each lobulus divided in two subsets of lobules. Although the lobular regions appeared to be a part of the cap-like area, it becomes apparent that these regions were connected to the disc-like area, when examining more ventral sections of the PPN (see. Fig. 6A, B, and C). The lobular structures merged ventrally into the disc-like area of the PPN. The large lobulus area, which linked the PPN and the longitudinal tracts, consisted of lung-shaped overlapping lobules, which were situated on a peduncle-like substructure (see. Fig. 6A and B).

The large lobulus area was situated anteriorly to the pectine neuropils and connected the subesophageal ganglion mass and the pectine neuropils (see. Fig. 7).

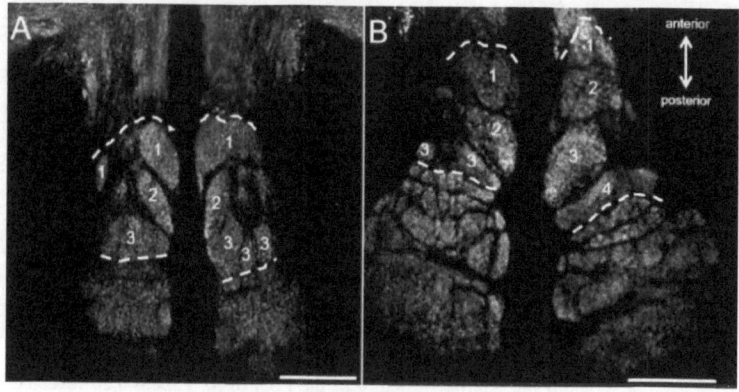

Figure 7: Sections of anti-Synapsin immunolabeling of the large lobulus area of *M. eupeus* T10 **(A)** and T22 **(B)**. Horizontal view, synaptic tissue in grey, ganglion midline shown by vertical dashed line **(A)**. The large lobulus area on each body hemisphere (limited by dashed white lines) consisted of up to four symmetrical lung-shaped lobules, shown by numbers in **(A)** and **(B)**. **(B)** Lobules appeared to imbricate each other. Smallest lobule indicated by number 1, largest by number 4. Scale bars: 100 μm.

The large lobulus area was 119.02 μm ± 5.651 in width and 176.13 μm ±13.069 in length (n=7). It extended to the subesophageal ganglion in posterior direction, connecting the posterior pectine neuropil and the longitudinal tracts within the merged subesophageal ganglion mass. The large lobulus area was composed of up to four lobules, which appeared lung-shaped and overlapping each other. Lobules appeared smaller in anterior direction and wider near the cap-like area. Although the cap-like area and the disc-like area exhibited roughly a glomerular and lobular architecture, the lobules of the large lobulus area appeared smooth and not glomerular (see. Fig. 7). The core of the large lobulus area appeared to be a stem-like structure, consisting of axonal bundles (see. Fig. 6).

3.3 Projection areas of sensory afferents

By combining the backfill technique and the immunohistological staining of the neuropil areas, it became possible to determine and analyze the projection areas of the pectinal teeth in the pectine neuropil (see Fig. 8). The BF of the distal pectinal tooth of the left pectine and the proximal pectinal tooth of the right pectine showed different localizations of projection areas within the respective neuropils. Whereas the sensory afferents of the distal pectinal tooth innervated the lateral areas of one PPN, the afferents of the proximal pectinal tooth projected into medial areas. One pectinal tooth had a distinctive ipsilateral projection region and never projected contralaterally. Lateral tracts proceeded along the left outer margin of the PPN and the large lobulus area of the left and right body hemisphere. The tracts at the level of the PPN enveloped part of it and fibers entered neuropilar areas from the outside of the neuropil. Afferents of the proximal pectinal tooth innervated the medial part of the right PPN and left through lateral tracts, which proceeded along the margin of the large lobulus area of the right pectine neuropil in anterior direction. The projection area of the proximal peg was visibly smaller compared to the projection area of the distal peg. Furthermore, no lateral tracts were visible at the level of the neuropil, only at the level of the large lobulus area.

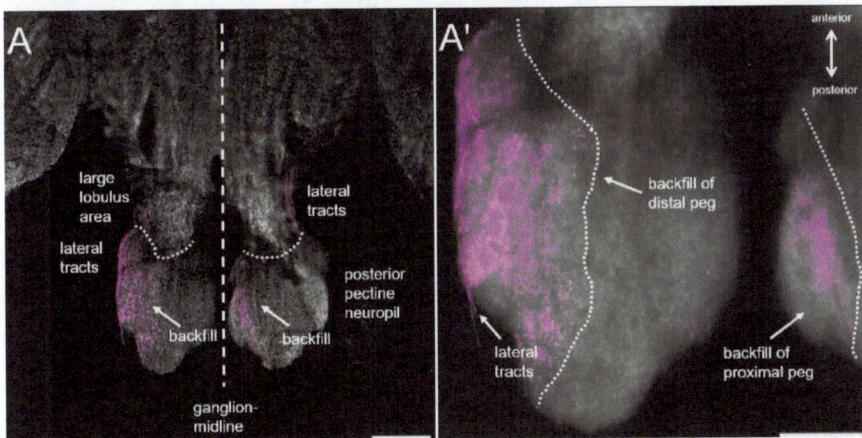

Figure 8: Neurobiotin™ backfill of distal pectinal tooth of the left pectine and proximal pectinal tooth of the right pectine of T10. Horizontal view. BF is shown in magenta, neuropilar regions in grey. **(A)** Projection of afferents of the distal peg in the left neuropil and of the proximal peg in the right pectine neuropil. Innervation areas of distal peg were limited to the lateral side of the left pectine neuropil, whereas projections of the proximal peg were only visible on the medial regions of the right neuropil. Lateral tracts proceeded along the left outer margin of the left pectine neuropil and the right outer margin of the large lobulus area of the right pectine neuropil. Large lobulus area was innervated by distal and proximal teeth. Vertical dashed line indicates ganglion midline. Scale bar: 100 µm. **(A')** Inset of **(A)**. Projection area of distal peg showed no innervation on the medial side of the left neuropil. Sensory afferents of the proximal peg innervated only a small medial area at the left margin of the right pectine neuropil. Scale bar: 50 µm

With the structural changes of the PPN from ventral to dorsal sections, a different innervation pattern became prominent, which can be seen in Fig. 9. The innervation by the distal peg showed a distinct localization pattern. Only the lateral part of the left neuropil exhibited labeled afferents, whereas the medial side did not. Both the lobules of the disc-like area and the cap-like area were innervated by these fibers. Furthermore, two glomeruli of the third row of the bell-shaped lamellar structure in the cap-like area showed innervation of afferents of the distal peg (see Fig. 9A and B). Lateral tracts proceeded along the left outer margin at the level of the cap-like structure and ended at the level of the adjacent located large lobulus area.

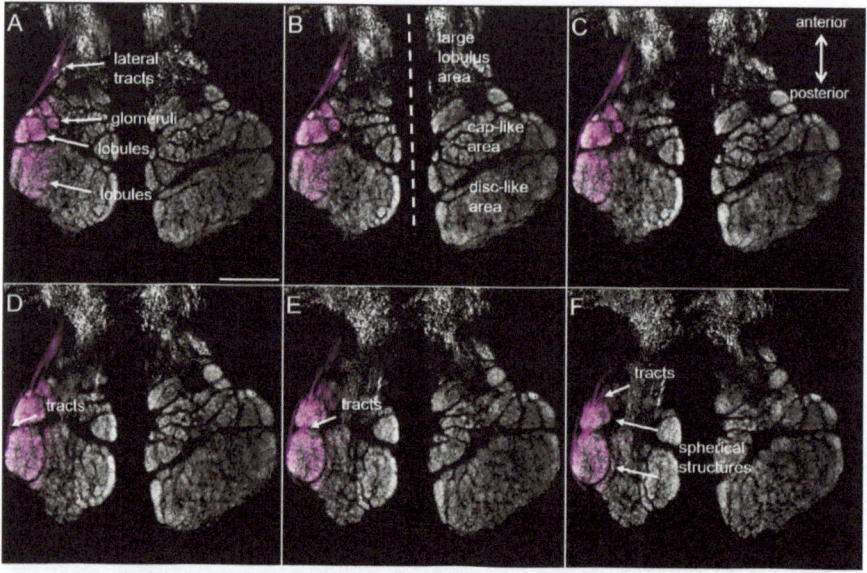

Figure 9: Projection areas of labeled sensory afferents of distal peg in the left posterior pectine neuropil of *M. eupeus* T22. Montage of 6 sections taken at 15 μm intervals through the pectine neuropil. **(A-F)** ventral to dorsal sections. Pectinal afferents of the distal peg were treated with Neurobiotin™ and labeled with Cy3™-Streptavidin (magenta). Neuropilar regions in grey. Ganglion midline indicated by dashed line in **(B)**. The neuropilar structures changed visibly from a distinct cap-like area and a disc-like area **(A-C)** to two spherical structures on the medial and lateral side, respectively. Additionally, more tracts became apparent **(D-F)**. Scale bar:100 μm

The internal structure, as well as the innervation pattern of the distal pectinal tooth of the left pectine changed in dorsal sections of the PPN. In ventral sections, the cap-like area appeared to contain two or three lateral lobular structures, which were divided into smaller lobules (see also Fig. 9A). In dorsal sections (approx. 90 µm in depth) the lateral areas of the cap-like area merged into smooth spherical regions with little fragmentation (see Fig. 9E and F). This transformation was also observed in the disc-like area and its palisade-shaped lobular architecture. At the dorsal end of the PPN, two distinct lateral lobes on each body hemisphere were visible.

The innervation pattern of the labeled distal peg changed from innervating one side of the left pectine neuropil, to a restricted projection area within the lateral lobules at the dorsal end of the PPN. Additionally, the tracts, which proceeded along the left lateral margin of the pectine neuropil, became more apparent within the pectine neuropil and went from the posterior part of the pectine neuropil to the anterior part (see. Fig. 9D-F).

Besides backfills of distal and proximal pectinal tooth, backfills of medial pegs were conducted. The following panel helps to compare the projection areas of these three differently located pectinal teeth (see Fig. 10)

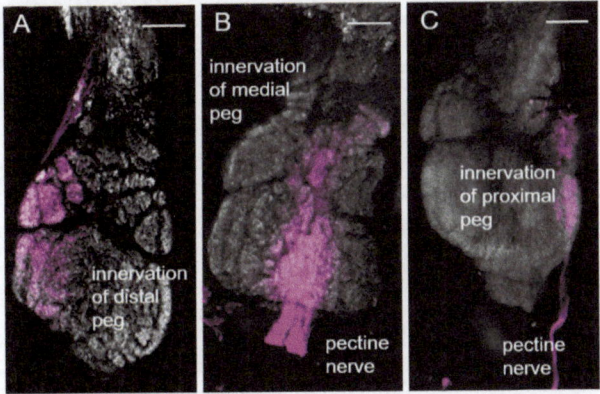

Figure 10: Sensory projection areas of a distal, medial and proximal pectinal tooth in the posterior pectine neuropil of animals T22, T31 and T33, respectively. Pectinal afferents of the distal peg, medial peg and proximal peg were treated with Neurobiotin™ and labeled with Cy3™-Streptavidin (magenta). Neuropilar regions in grey. Left posterior pectine neuropils of the respective animals are shown. **(A)** Innervation of the pectine neuropil by a distal backfilled pectinal tooth (section taken at 140 µm depth from the ventral side). **(B)** Sensory fibers of the medial peg innervated wedge-shaped centrally located neuropil areas. Pectine nerve entered posteriorly from the ventral side (section taken at 140 µm depth from the ventral side). **(C)** The proximal peg projected into medial areas near the ganglion midline (not shown). Pectinal nerve entered posteriorly from the ventromedial side (section taken at 140 µm depth from the ventral side). Scale bars: 50 µm

Distally located pegs exhibited a distinct projection area at the lateral side of the left pectine neuropil. Similar limited projection areas were visible for the proximal and the medial pegs. The afferent fibers of the proximal peg of T31 were bundled within the pectine nerve, entered the PPN from the ventral side and innervated medial neuropil areas near the ganglion midline (not shown) within the left pectine neuropil (see Fig. 10C). Labeled afferents of a medial pectinal tooth of T33, entered the PPN through fibers, which were situated ventro-caudally and projected in centrally situated neuropilar areas within the PPN. The innervation pattern of the medial peg was wedge-shaped; wider at the posterior end of the neuropil and smaller at the anterior end. Lateral and medial areas were not innervated by the afferents of the medial peg.

Not only the posterior pectine neuropil received sensory input from the pectinal teeth, but also the adjacent located large lobulus area, which connected the PPN and the synganglion (see. Fig. 11).

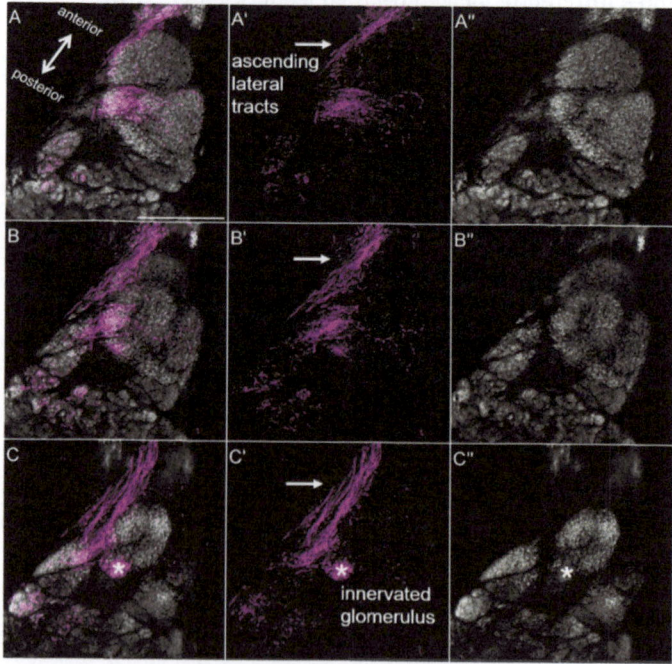

Figure 11: Neurobiotin™ backfill of distal peg of the left pectine of animal T7. Projection area of afferents within the LLA. Horizontal view. Ventral to dorsal sections. Backfill (BF) is shown in magenta. **(A, B, C)** merge of BF (magenta) and synaptic tissue (grey). **(A', B', C')** Ascending lateral tracts and one innervated glomerulus in **(C')**. **(A'', B'', C'')** Synaptic tissue of the LLA. Scale bar: 50 μm.

After leaving the ovoid shaped pectine neuropil through lateral tracts, the sensory fibers of the distal pectinal tooth innervated the large lobulus area. The innervation proceeded from lateral located tracts to medial regions within the lobules of the large lobulus area. In ventral sections, the projection areas within one lobulus showed a diffuse innervation pattern, whereas in dorsal section, the innervation appeared more restricted to one glomerulus. The tracts became more visible and thicker in dorsal sections of the large lobulus area and projected into one distinct glomerulus in close vicinity of the tracts (see Fig. 11C). Tracts, which left the large lobulus area in anterior direction, projected near the ganglion midline into longitudinal tracts.

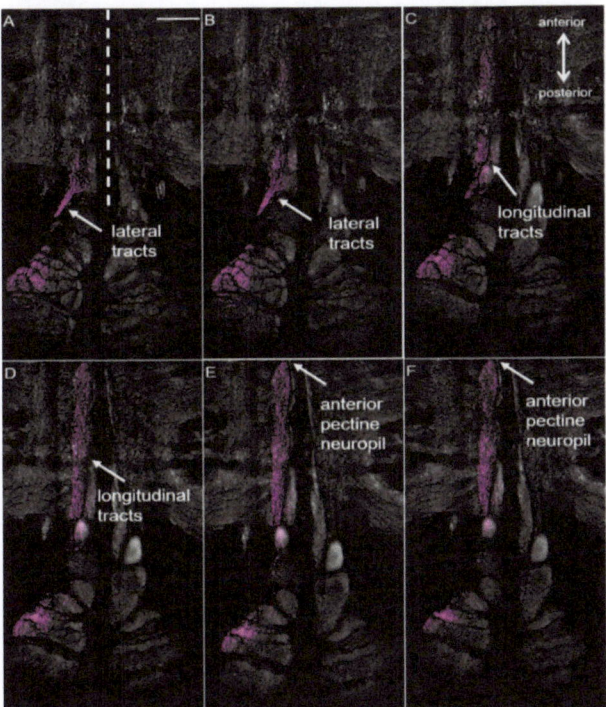

Figure 12: Projection areas of sensory afferents of distal peg in the left PPN of *M. eupeus* T22. Montage of 6 sections taken at 10 μm intervals through the subesophageal ganglion mass. **(A-F)** Ventral to dorsal sections. Pectinal afferents of distal peg were treated with Neurobiotin™ and labeled with Cy3™-Streptavidin (magenta). Neuropilar regions in grey. Ganglion midline indicated by dashed white line in **(A)**. Lateral tracts shown by white arrows in **(A)** and **(B)**. Longitudinal tracts near the ganglion midline indicated by white arrows in **(C)** and **(D)**. Anterior pectine neuropil specified by white arrows in **(E)** and **(F)** at the level of the third walking leg neuromere. Scale bar: 100 μm.

The innervation of one peg within the pectinal system showed a large network of involved neuropilar structures, which were not only limited to the posterior pectine neuropil, but also present in the subesophageal ganglion mass (see Fig. 12). Lateral tracts, which connected the projection areas of the PPN and the anterior pectine neuropil, were visible at a section approx. 140 μm within the pectine neuropil. They extended anteriorly and merged into the longitudinal tracts in dorsal sections. The mean length of the longitudinal tracts was 517.65 μm ± 6.461 (n=4). At the anterior end of the longitudinal tracts projected into the anterior pectine neuropil, which was cone-shaped and exhibited no fragmentation.

4 Discussion

Previous studies on the scorpion pectines and pectinal network within the nervous system were able to clarify some important questions regarding morphology (Gaffin, et al., 1997; Wolf, 2008), function (Krapf, 1986; Gaffin, et al. 1997; Melville, 2003; Mineo, 2006; Kladt, Wolf and Heinzel, 2007), response properties (Gaffin et al. 1997; Gaffin, 2002), innervation (Wolf, 2008) and approximate projection areas (Wolf, 2008). However, nobody could fully answer the question on how sensory afferents from the receptors of single pegs innervate the pectine neuropil and which other parts of the CNS receive input from the pectines. In this study, I was able to answer these questions and showed the topographic innervation pattern of chemo- and mechanosensory fibers from single pegs in the posterior pectine neuropil network.

4.1 Discussion of methods

In the beginning of this project, a suitable method had to be developed and adjusted for backfilling sensory fibers of single pectinal teeth. Problems arose in a couple of steps within the protocol and had to be eliminated. The fixation method of the animals during the backfill had to be adjusted: instead of an invasive method, which involved pinning the animals to the cork plate, a noninvasive method with plasticine was used. This ensured the wellbeing of the animals during the backfill procedure, which was also adjusted from one day to three days. Additionally, the embedding method was altered, from gelatin/albumin to agarose, because the specimen was more visible in the embedding medium and facilitated further processing. But despite careful and thoughtful implementation of improvements, some of the experiments have not show any positive results. Therefore, the priority was to adjust and improve the methods, before starting with the immunohistochemistry. For this reason, the first experiments had no immunocytochemical treatment and only DAPI served as counter staining. In this case, only the approximate pectine area could be determined, but no structure of the neuropil itself.

© Springer Fachmedien Wiesbaden GmbH, part of Springer Nature 2019
D. Drozd, *Topographic Organization of the Pectine Neuropils in Scorpions*,
BestMasters, https://doi.org/10.1007/978-3-658-25155-0_4

The results of the first experiments, showed a general idea of the localization of projection areas within the respective posterior pectine neuropil. Axonal fibers of distal and proximal located pegs entered ventrally the synganglion and projected in spatially different areas of the PPN (see Fig. 3), which was also shown by Wolf in 2008. This implies a topographically structured pectine neuropil (Brownell, 1998), which is unusual in chemosensory systems, but not unique.

These results gave only little information about the pectine neuropil system. To gather additionally information about the morphology of the pectine neuropil and consequently also the innervation of specific substructures, the synaptic marker anti-SYNORF1 was used. This allowed a holistical view onto the system and helped to examine the interaction between different components of the pectinal network.

4.2 Analysis of the architecture of the posterior pectine neuropil

Works of previous studies (Brownell, 1998; Melville, 2000; Wolf, 2008; Gaffin, 1997;) started to unravel the architecture of the pectine neuropils, by backfilling the pectines or sensilla with Cobalt/Nickel, by histology and later by immunocytochemistry. However, the results were partially limited by the applied techniques and methods. But nevertheless, the results gave the first insights in the chemo- and mechanosensory processing in scorpions and the associating neuropilar structures within the synganglion (see Introduction).

In this study, single pegs in varying proximity to the base, were backfilled with the tracer Neurobiotin™ and visualized by using Streptavidin™-Cy3. Anti-SYNORF1 served as a synaptic marker, which labeled neuropilar regions and allowed to characterize the pectinal sensory network. Dark appearing boundaries around specific subdivisions where used to determine the glomerular or lobular structures.

4.2.1   The structure of the disc-like area

Horizontal sections revealed the extent of the pectinal sensory network; beginning in the most posterior part of the merged subesophageal ganglion mass and ending at the level of the third walking leg neuromere (see Fig. 5). The bilateral pectine neuropil appeared rounded in shape at the posterior end, which bears similarity to the Blumenthal neuropil of the spider *Cupiennius salei*, because of its rounded shape and rough glomerular structure. However, this neuropil within the synganglion mass of the spider, processes sensory information from hygro-

and thermoreceptors on the tarsal organs (Anton and Tychi, 1994) and is situated in the center of the subesophageal ganglion mass. Each of the pectine neuropils was divided into two distinct hemispheres, which was also stated by Melville, 2000. He labeled horizontal sections of the subesophageal ganglion of *Paruroctonus mesaensis* also with anti-SYNORF1 and analyzed the pectine neuropils regarding the cytoarchitecture of the pectinal neuropil and projection areas of bimodal receptors and hair sensilla on the pectines. In his study, he called the posterior part of the neuropil the disc and the anterior part the cap, which had a fibrous outer cortex and a medullar region. His results suggested that the disc was considerably smaller than the cap, which could neither be verified by Wolf, 2008, nor in this recent study. In horizontal sections of *M. eupeus* the border between these two parts of the neuropil, seemed to be horizontal and slightly arcuate. In parasagittal and frontal sections of *Buthus occitanus* and *Vaejovis spinigerus*, this border seemed slanting (Drozd, 2014; Hüll, 2014; Wolf, 2017). Differences regarding size of specific substructures within the pectine neuropil could be due to analyzing different species. Here in this study the old-world scorpion *M. eupeus* of the Buthidae served as test species and in Melville's study *P. mesaensis*, a member of the Vaejovidae was used. These scorpions are phylogenetically and spatially distant to each other and therefore bear differences in some morphological features, regarding the numbers of pegs on a pectine, the sensilla, consequently also the structure of the respective receptors and possibly also a different architecture of the pectine neuropils (Wolf, 2008; Sharma, 2015). This potential variation of the neuropil structure and its functional consequences has yet to be analyzed in further studies. Nevertheless, both old-world and new-world scorpions show similarities in crucial details, concerning the basic functional and anatomical pathways in chemo-and mechanosensory processing, showing that the differences have meaning in neurophylogenetic studies, at least to some extent (Ivanov and Balashov, 1979; Foelix and Müller-Vorholt, 1983). The disc-like area in the posterior part of the pectine neuropil of *Mesobuthus eupeus* was notably larger than suggested by Melville and consisted of rough lamellar lobuli, showing almost no structural distinction and glia cell boundaries (Melville, 2000). These lobuli were comparable to mechanosensory neuropilar structures found in the insect brain (Hanström, 1928; Homberg, Christensen, Hildebrand, 1989) and malacostracan crustacean brain (Schmidt and Ache, 1992; Sandeman et. al. 1992; Fanenbruck and Harzsch, 2005; Schachtner, Schmidt and Homberg, 2005; Harzsch and Hansson, 2008; Strausfeld, 2009) or in the corpus lamellosum of the myriapod brain (Hörberg, 1931; Fahlander 1938; Sombke, Harzsch and Hansson, 2011; Sombke, Rosenberg, Hilken, 2011; Sombke et. al., 2012), due to their laminar, palisade-shaped structure. Mechanosensory neuropils in the arthropod brains are usually elongated, sausage-shaped and arranged in a palisade-like-fashion (Foelix and Müller-Vorholt, 1983; Loesel, Nässel and Strausfeld 2002; Strausfeld, 2009). This can be seen in Fig. 10A and B in the posterior part of the neuropil (disc-like area). Wolf suggested, the whole posterior pectine neuropil to code for chemosensory information (Wolf, 2017). If the disc integrates only chemosensory or mechanosensory information, can not be stated yet, due to a lack of data regarding the innervation by the respective afferents. Both chemo- and mechanosensory information are perceived by the pectines and therefore both should be processed in the pectine neuropil (Wolf, 2008, 2017).

Another interesting observation could be seen in Fig. 6, where the disc-like area was merged with the lobular areas of the cap-like area. In ventral sections of the pectine neuropil, two distinct lobular layers were visible in the disc-like area (see Fig. 6A, B), which separated in further dorsal sections into two bilateral lobules in the cap-like area in each pectine neuropil, respectively. This means these two hemispheres are no morphological separated units, but are connected in a specific manner. This observation was not mentioned neither by Melville, Wolf, Gaffin nor Brownell (Brownell, 1988, 1998; Gaffin and Brownell, 1997, 2002, 2010; Melville, 2000; Wolf, 2008, 2017) and the meaning and consequently the function of these structures has yet to be analyzed and interpreted.

4.2.2 The structure of the cap-like area and the large lobulus area

The superstructure of the cap consisted of three arcuate layers, which in turn contained smaller glomeruli (see Fig. 5 and 6). These glomeruli were continuous and not distinguishable, therefore the number of the glomeruli couldn't be counted. Similar observations were made by Brownell, 1998. He showed for the first time an overall overview of the posterior pectine neuropil (Brownell, 1998). In presumably transversal sections of the synganglion, he stated that the chemosensory neuropil in *Paruroctonus* (Family: Vaejovidae), resembled that of the mandibulate arthropod antennal lobe, by means of its glomerular architecture. But he also stated, that the pectine neuropil consisted of microglomeruli which were arranged in concentric layers. This finding was verified by the works of Wolf in 2008, who also worked with scorpions from the family of the Vaejovidae (Wolf, 2008). Wolf conducted backfills and histology on leg and pectine afferent arborizations and analyzed the central nervous projections. Axons from the mechano- and chemosensory receptor cells entered the posterior pectine neuropil, where they projected into a layered neuropil with a rough glomerular architecture. Wolf called the superstructure of the neuropil 'onion-peel'-like, because the neuropil consisted of arcuate lobular neuropil layers, which appeared to be smaller in the center and wider in the outer cortex of the neuropil. Furthermore, previous studies showed that there are up to 140 glomerular structures in the pectine neuropil of *Buthus occitanus*, within 19 layers (Drozd, 2014, Hüll, 2014). However, these 'glomeruli' were not clearly separated, but rather continuous. Results of *M. eupeus* suggest a rougher glomerular structure of the arcuate layers and are close to the results by Wolf, with the difference that *M. eupeus* exhibited only three of these arcuate glomerular layers. Most notably was the existence of two lateral lobules on each side of the posterior pectine neuropil. The meaning and the function of this interesting area has yet to be studied.

Glomeruli or glomerular structures are generally associated with olfaction and chemosensory processing (Strausfeld and Hildebrand, 1999; Eisthen, 2002). In arthropods, molluscs, nematodes and vertebrates brain regions for olfactory input share many features, which are intriguingly similar and therefore propose a homolog evolution of chemosensory systems (Eisthen, 2002; Schachtner, Schmidt and Homberg, 2005; Sombke et. al., 2011). One of these similar features are the glomerular structured olfactory neuropils (Kosaka et al., 1998; Wachowiak and Shipley, 2006). In arthropods, glomeruli are often spherical structures in the olfactory pathway, usually located in the deutocerebrum, which is associated with the respective olfactory sensory organ, like the antennae or the maxillary palps in different taxa (Strausfeld and Hildebrand, 1999; Galizia and Rössler, 2010; Hansson and Stensmyr, 2011; Sombke, Rosenberg and Hilken, 2011; Li and Stephen, 2015). Projection neurons send their axon terminals from the glomeruli into the higher brain regions, namely the mushroom bodies and the lateral horn in the protocerebrum, where the chemosensory and other sensory information is processed and integrated. The glomeruli which are found in scorpions, share similarities to the glomeruli found in other arthropod species, but are also distinctively different. Scorpions have relatively small glomeruli (microglomeruli), which are not arranged in a lobe, but in concentric layers around the peduncle of the large lobulus area. Also, the number of glomeruli seems variable and not easy to properly quantify in scorpions, whereas in some other arthropods e.g. in *Drosophila melanogaster* (Stocker et al., 1990) it is measurable and the specific ORN's (olfactory receptor neuron), which project into one certain glomerulus are known (Eisthen, 2002; Hansson and Stensmyr, 2011). This particular feature of receptor specific organization within the olfactory system could not be verified in the previous studies on scorpions.

The next relay station in the pectine system was the large lobulus area, which was adjacent to the posterior part of the rounded pectine neuropil (see Fig. 6A, B and Fig. 7). This neuropilar structure exhibited different features as the previous mentioned posterior pectine neuropil, as there were no glomerular structures and it was not shaped like the typical pectine neuropil (Brownell, 1988, 1998; Melville, 2000; Drozd, 2014; Wolf, 2008, 2017). The neuropilar structure resembled a pine cone with overlapping lobular neuropils, which were positioned on a peduncle (see. Fig. 6A, B). It's not clear yet, what exact function this area has and if only one sensory modality is processed within. This question could not be answered and requires further studies.

4.2.3 Somatotopic innervation of the pectine neuropil

The disc-like area was the first integration center in the sensory pathway. Arborizations of the pectine nerve enclosed the pectine neuropil, entered it posteriorly from the ventral side and projected into the disc-like area, cap-like area and the large lobulus area. According to Wolf, roughly 140.000 mechano- and chemosensory sensory afferents in males and 80.000 in females enter the posterior pectine neuropil in *Vaejovis spinigerus* (Wolf, 2008). But comparable data is lacking in this study. In both, the posterior and anterior part of the pectine neuropil, a distinct somatotopic innervation pattern was observed. The backfill analysis showed that distally located pectinal teeth innervated the outer lateral part of the pectine neuropil (see Fig. 9, 10, 11), whereas proximal located peg afferents projected into the medial parts (near the ganglion midline) of the pectine neuropil. Afferents of pegs located medially on the pectine spine exhibited a central projection pattern within the pectine neuropil. This observation was verified by Wolf in 2008, however in reversed order. This might be due to errors in the interpretation (pers. comm. Harald Wolf). Also, the large lobulus area showed a peg specific projection pattern, with innervated glomeruli associated to the lobules (see Fig. 12). If this structure processes mechano- or chemosensory input, is still unclear. Melville suggested this area to process only mechanosensory input (Melville, 2000), but it is not exactly clear, if both mentioned areas are really the same. The results of the backfill analysis propose a somatotopic, instead of a chemotopic organization within the pectine neuropil (Brownell, 1998; Wolf, 2008, 2017), which was also observed in the solpugids' malleolar system (Brownell, 1974, 1998), but not particularly in other arachnids or arthropods so far. Chemosensory information is usually processed chemotopically in dense glomeruli (Homberg et al., 1989; Sombke, Harzsch and Hansson, 2011; Sombke et. al., 2012; Loesel et al., 2013) in the antennal or olfactory lobes of Mandibulata

(Sombke, Harzsch and Hansson, 2011), where the information of the ORN's is conveyed to the respective glomerulus (Eisthen, 2002), rather than topographically. Different receptors neurons innervate one specific glomerulus (Vosshall and Stocker, 2007). This allows spatial segregation of information from chemoreceptors, which respond to different chemicals. After leaving the antennal lobes, the information is projected via interneurons to higher brain regions (mushroom bodies or hemiellipsoid bodies) in the protocerebrum (Homberg et al., 1989; Strausfeld et al., 1998). Interestingly, longitudinal tracts left the posterior pectine neuropil and the large lobulus area laterally and proceeded along the ganglion midline to the anterior pectine neuropil at the level of the third walking leg neuromeres (see. Fig. 13). An analysis if these tracts project into the mushroom bodies in the protocerebrum of the scorpion is still missing, but it would be interesting to know, if there are any resemblances to the mandibulate brain. In contrast, mechanosensory input is processed in a somatotopic fashion in the arthropod brain (Anton and Barth, 1993). This is an important feature to process external information, by its occurrence in the vicinity of the animal. A spatially high resolution of mechanosensory input and its processing thus leads to a higher number of behavioral modes. Since both chemo- and mechanosensory information are perceived by the pectines, both have to be processed in the pectine neuropil as a consequence. Melville showed in 2000, that hair sensilla on the pectine project into the outer cortex area within the cap-like area (Melville, 2000). In comparison to this present study, there was no visible cortex area, but a spherical lobular area within the cap-like area (see Fig. 9A). When backfilling the afferents of either a distal or proximal peg, there was always innervation of this lobular area, which makes sense, because each peg bears hair sensilla and peg sensilla. This suggests that the lobular areas are presumably mechanosensory regions within the pectine neuropil. Besides his study also showed the innervation of afferents of bimodal receptors within a medullar region, which could be equivalent to the layered glomerular region in the cap-like area. The comparison would propose that both types of information are processed in spatially different regions of one distinct projection area within the pectine neuropil. It would be convenient if this extremely heterogeneous neuropil would not only process chemosensory input in one distinct region (e.g. in the layered glomerular region within the cap), but also, depending on the location of the peg in one particular segment of this region (e.g. in the lateral part of the layered glomerular region within the cap). But if hair sensilla or peg sensilla of one specific located peg innervate only one explicit region within the neuropil could not be verified in this study. Furthermore, different animals were used (as mentioned above), which could lead to false assumptions and misinterpretations.

Somatotopic processing of olfactory information in the arthropod brain was not often observed. As mentioned above, in the solpugids' chemosensory malleolar system a similar situation could be identified (Brownell, 1974, 1998), but with no explicit explanation on how the data were gathered. In addition, the American cockroach *Periplanata americana* and the moth *Manduca sexta* showed different response areas in the macroglomerular region in the antennal lobe, when stimulating the antenna with pheromones on different locations (Hösl, 1990; Heinbockel and Hildebrand, 1998). Normally, temporal and chemotopic resolution take up a more important position in the arthropod brain, because arthropod antennae are usually too small or slender to process spatially relevant chemosensory information in aqueous or aerial media (Wolf, 2017). For scorpions this situation is different, as the pectines are close to the substrate and the peg sensilla can perceive more odour molecules. This might lead to the fact, that olfactory information can be resolved spatially in the nervous system by a somatotopic organization of the afferent projections within the pectine neuropil. Chemical structures or gradients on the substrate can therefore be perceived and processed, leading to a higher chance for kin recognition, mate trailing or prey capture.

5 Conclusion

The results of this study showed new insights into the architecture of the scorpions pectine neuropil and the somatotopic innervation pattern by afferent neurons of differently located pegs. Moreover, these results were compared and discussed with previous studies, which showed a more comprising view of the chemo- and mechanosensory processing strategy in scorpions. There are some similarities between the primary chemosensory neuropils in mandibulates (Sombke, Rosenberg and Hilken, 2011; Loesel, 2013) and the scorpion pectines (Wolf, 2008, 2017; Wolf and Harzsch, 2012). In mandibulates, there are two primary neuropils: one for chemosensory information and one for mechanosensory information. Both of these neuropils are adjacent and differently structured. The antennal or olfactory lobes are comprised of distinct glomeruli, whereas the antennal mechanosensory neuropils (e.g. antennal mechanosensory motor center) are rather palisade-shaped or elongated (Hanström, 1928; Hörberg, 1931; Fahlander 1938; Homberg et al., 1989; Schmidt and Ache, 1992; Sandeman et al., 1992; Fanenbruck and Harzsch 2005; Harzsch and Hansson, 2008; Sombke, Harzsch and Hansson, 2011; Sombke, Rosenberg, Hilken, 2011). A similar situation is present in the scorpion pectine neuropil. The neuropil consists of different subdivisions, which are glomerular or lobular. The similarities of the chemosensory neuropils found in arthropod nervous system, might propose the idea of a basic plesiomorphic trait within this taxon, rather than a synapomorphy of the Mandibulata. But despite these similarities, there are also differences, which includes the completely different location of the primary chemosensory organs. Mandibulates bear their antennae on the head segments, whereas scorpions have ventrolateral appendages, as the primary chemosensitive organs. Furthermore, the overall structure of the primary chemo- and mechanosensory neuropils in Mandibulata are different than in scorpions. Scorpions do not have distinct spherical glomeruli, but rather flattened continuous glomeruli, which are organized in arcuate layers. Finally, the antennal lobe of the mandibulates do not show a somatotopic organization of chemosensory input (except the previous mentioned examples), but a chemoreceptor type specific innervation, as well as a temporal resolution of the information in the brain, whereas the scorpion pectine neuropil exhibits a topographic mapping of chemosensory input.

5.1 Outlook

The pectine neuropil of the scorpion is an interesting study object, as there are many questions left to answer. Answering these questions, might give new approaches for phylogenetic studies of the evolution of arthropods. Some of these questions are:

• In which regions are mechano- and chemosensory input processed respectively?

• Is the innervation of peg afferents from different located pegs overlapping or are there distinct limitations?

• What is the exact function of the large lobulus area?

• Is there a connection to higher brain regions, such as the mushroom bodies or the arcuate body?

References

Abushama F. (1964) On the behaviour and sensory physiology of the scorpion *Leiurus quinquestriatus*. Animal Behavior, 12, pp. 140-153.

Alexander A.J. (1957) The courtship and mating of the scorpion, *Opisthophthalmus latimanus*. Journal of Zoology, 128, pp. 529–544.

Alexander A.J. (1959) Courtship and mating in the buthid scorpions. Journal of Zoology, 133, pp. 145–169.

Anton S. and Tichy H. (1994) Hygro- and thermoreceptors in tip-pore sensilla of the tarsal organ of the spider *Cupiennius salei*: innervation and central projection. Cell and Tissue Research, 278, pp. 399-407.

Anton S. and Barth F.G. (1993) Central nervous projection patterns of trichobothria and other cuticular sensilla in the wandering spider *Cupiennius salei* (Arachnida, Araneae). Zoomorphology, 113, pp. 21-32.

Babu K.S. (1965) Anatomy of the central nervous system of arachnids. Zoologische Jahrbücher Anatomie, 82, pp. 1-154.

Babu K.S. and Barth F.G. (1984) Neuroanatomy of the central nervous system of the wandering spider, *Cupiennius salei* (Arachnida, Araneida). Zoomorphology, 104, pp. 344–359.

Babu K.S., Sreenivasulu K., Sekhar V. (1993) Sensory projections of identified coxal hair sensilla of the scorpion *Heterometrus fulvipes* (Scorpionidae). Journal of Bioscience, 18, pp. 247-259.

Barth F.G. (2004) Spider mechanoreceptors. Current Opinion in Neurobiology, 14, pp. 415-422.

Boeck J. and Tolbert L.P. (1993) Synaptic organization and development of the antennal lobe in insects. Microscopy Research & Technique, 24, pp. 260-280.

Boyden B.H. (1978) Substrate selection by *Paruroctonus boreus* (Girard) (Scorpionida: Vejovidae). Master's Thesis, Idaho State University, Pocatello, ID.

© Springer Fachmedien Wiesbaden GmbH, part of Springer Nature 2019
D. Drozd, *Topographic Organization of the Pectine Neuropils in Scorpions*,
BestMasters, https://doi.org/10.1007/978-3-658-25155-0

Bristowe W.S. (1958) World of Spiders. Collins New Naturalist, ISBN 0-8008-8598-8. Nov 1958, pp. 304-350.

Brownell P.H. and Farley R.D. (1974) The organization of the malleolar sensory system in the solpugid, *Chambria* Sp. Tissue and Cell, 6, pp. 471-485.

Brownell P.H. (1988) Properties and functions of the pectine chemosensory system of scorpions. Chemical Senses, 13, pp. 557.

Brownell P.H. (1998) Glomerular Cytoarchitectures in Chemosensory Systems of Arachnids. Annals of the New York Academy of Sciences, 855, pp. 502-507.

Butterfield N.J. (2003) Exceptional fossil preservation and the Cambrian Explosion. Integrative and Comparative Biology, 43, pp. 166–177.

Carthy J.D. (1966) Fine structure and function of the sensory pegs on the scorpion pecten. Experientia, 22, pp. 89-91.

Cutler B. (1980) Arthropod cuticle features and arthropod monophyly. Cellular and Molecular Life Sciences, 36, pp. 953.

de Brito Sanchez M.G., Lorenzo E., Songkung S., Liu F., Giurfa M. (2014) The tarsal taste of honey bees: behavioral and electrophysiological analyses. Frontiers in Behavior Neuroscience, 8, pp. 25.

Dethier V.G. (1947) Chemical Insect Attractants and Repellents. Blakiston, Philadelphia, Annual review in Entomology, 1, pp. 181-202.

Dunlop J.A. and Braddy S.A. (2001) Scorpions and their sister-group relationships. V. Fet, P.A. Selden (Eds.), Scorpions 2001 – In Memoriam Gary A. Polis, British Arachnological Society, pp. 1-24.

Dunlop J.A. and Webster M. (1999) Fossil evidence, terrestrialization and arachnid phylogeny. The Journal of Arachnology, 27, pp. 86-93.

Dumpert K. (1978) Spider odor receptor: Electrophysiological proof. Experientia, 34, pp. 754-756.

Drozd D. (2014) Das primäre chemosensorische Neuropil der Skorpione – Analyse der "glomerulären" Struktur, Teil 1. Bachelor of Science Thesis in Biology, pp. 1-26.

Dzik J. (2007) The Verdun Syndrome: simultaneous origin of protective armour and infaunal shelters at the Precambrian–Cambrian transition, in Vickers-Rich, Patricia; Komarower, Patricia, The Rise and Fall of the Ediacaran Biota, 286, London: Geological Society, pp. 405–414.

Edgecombe R.S. and Murdock L.L. (1992) Central projections of axons from taste hairs on the labellum and tarsi of the blowfly, *Phormia regina* Meigen. Journal of Comparative Neurology, 315, pp. 431-444.

Eisthen H.L. (2002) Why Are Olfactory Systems of Different Animals So Similar? Brain, Behavior and Evolution, 59, pp. 273-293.

Egan M.E. (1976) The chemosensory bases of host discrimination in a parasitic mite. Journal of Comparative Physiology, 109, pp. 69-89.

Fanenbruck M. and Harzsch S. (2005) A brain atlas of *Godzilliognomus frondosus* Yager, 1989 (Remipedia, Godzilliidae) and comparison with the brain of *Speleonectes tulumensis* Yager, 1987 (Remipedia, Speleonectidae): implications for arthropod relationships. Arthropod Structure & Development, 34, pp. 343-378.

Fahlander K. (1938) Beiträge zur Anatomie und systematischen Einteilung der Chilopoden. Zoologiska Bidrag från Uppsala, 17, pp. 1-148.

Farley R.D. (2001) Development of segments and appendages in embryos of the desert scorpion *Paruroctonus mesaensis* (Scorpiones: Vaejovidae). Journal of Morphology, 250, pp. 70-88.

Foelix R.F. (1970) Chemosensitive hairs in spiders. Journal of Morphology, 132, pp. 313-333.

Foelix R.F. and Chu-Wang I.W (1973) The morphology of spider sensilla I. mechanoreceptors. Tissue and Cell, 5, pp. 451-460.

Foelix R.F. and Müller-Vorholt G. (1983) The fine structure of scorpion sensory organs. II. Pecten sensilla. British Arachnological Society, 6, pp. 68-74.

Foelix R.F. and Schabronath J. (1983) The fine structure of scorpion sensory organs. I. Tarsal sensilla. British Arachnological Society, 6, pp. 53-67.

Foelix R.F. (1985) Mechano- and Chemoreceptive Sensilla. In: Barth F.G. (ed) Neurobiology of Arachnids. Springer, Berlin, Heidelberg. pp. 180-248.

Gaffin D.D. and Brownell P.H. (1992) Evidence of Chemical Signaling in the Sand Scorpion, *Paruroctonus mesaensis* (Scorpionida: Vaejovida). Ethology, 91, pp. 59-69.

Gaffin D.D. (1994) Chemosensory Physiology and Behavior of the Desert Sand Scorpion, *Paruroctonus Mesaensis.* Dissertation thesis. University of Oregon, pp. 1-181.

Gaffin D.D. and Brownell P.H. (1997) Response properties of chemosensory peg sensilla on the pectines of scorpions. Journal of Comparative Physiology A, 181, pp. 291-300.

Gaffin D.D. and Brownell P.H. (1997) Electrophysiological evidence of synaptic interactions within chemosensory sensilla of scorpion pectines. Journal of Comparative Physiology A, 181, pp. 301–307.

Gaffin D.D. (2002) Electrophysiological Analysis of Synaptic Interactions Within Peg Sensilla of Scorpion Pectines. Microscopy Research & Technique, 58, pp. 325–334.

Gaffin D.D. (2010) Analysis of sensory processing in scorpion peg sensilla. Journal of Arachnology, 38, pp. 1-8.

Galizia C.G. and Menzel R. (2000) Odour perception in honeybees: coding information in glomerular patterns. Current Opinion in Neurobiology, 10, pp. 504-510.

Galizia C.G and Menzel R. (2001) The role of glomeruli in the neural representation of odours: results from optical recording studies. Journal of Insect Physiology, 47, pp. 115-130.

Galizia C.G. and Rössler W. (2010) Parallel Olfactory Systems in Insects: Anatomy and Function. Annual Review of Entomology, 55, pp. 399-420.

Hanström B. (1923) Further notes on the central nervous system of arachnids: Scorpions, phalangids, and trap-door spiders. The Journal of Comparative Neurology, 35, pp. 249–274.

Hallberg E. and Hansson B.S. (1999) Arthropod sensilla: Morphology and phylogenetic considerations. Microscopy Research & Technique,47, pp. 428-439.

Hansson B.S. and Stensmyr M.C. (2011) Evolution of Insect Olfaction. Cell Press, Neuron Review, 72, pp. 698-711.

Harzsch S., Hansson B.S. (2008) Brain architecture in the terrestrial hermit crab *Coenobita clypeatus* (Anomura, Coenobitidae), a crustacean with a good aerial sense of smell. BMC Neuroscience, 9:58. doi:10.1186/1471-2202-9-58

Heinbockel, T and Hildebrand. J.G. (1998) Antennal receptive fields of phero-mone-responsive projection neurons in the antennal lobes of the male sphinx moth *Manduca sexta*. Journal of Comparative Physiology A, 183, pp. 121-133.

Hjelle J.T. (1990) Anatomy and morphology. The Biology of Scorpions. Stan-ford University, Stanford, pp. 9–63.

Homberg U., Christensen T.A., Hildebrand J.G. (1989) Structure and function of the deutocerebrum in insects. Annual Review of Entomology, 34, pp. 477-501.

Hörberg T. (1931) Studien über den komparativen Bau des Gehirns von *Scutigera coleoptrata*. L. Lunds Universitets Årsskrift N.F. Avd. 2, 27, pp. 1-24.

Hösl M. (1990) Pheromone-sensitive neurons in the deutocerebrum of *Peri-planeta americana*: receptive fields on the antenna. Journal of Compara-tive Physiology A, 167, pp. 321-327.

Hüll S.K. (2014) Das primäre chemosensorische Neuropil der Skorpione eine Analyse der "glomerulären" Struktur, Teil 2. Bachelor of Science Thesis in Biology, Ulm University, Germany

Ignell R. and Hansson B.S. (2005) Projection patterns of gustatory neurons in the suboesophageal ganglion and tritocerebrum of mosquitoes. The Journal of Comparative Neurology, 492, pp. 214-233.

Ivanov V.P. and Balashov Y.S. (1979) The structural and functional organization of the pectine in a scorpion *Buthus eupeus* Koch (Scorpiones, Buthidae) studied by electron microscopy. The fauna and Ecology of Arachnida, 85, Trudy Zoological Institute of Leningrad, pp. 73-87.

Kanzaki R. and Ikeda A. (1994) Morphology and Physiology of Pheromone-Triggered Flip-Flopping Descending Interneurons of the Male Silkworm Moth, *Bombyx mori*. In: Kurihara K., Suzuki N., Ogawa H. (eds) Sprin-ger, Tokyo. Olfaction and Taste XI. pp. 851-851.

Keil T.A. (1976) Sinnesorgane auf den Antennen von *Lithobius forficatus* L. (Myriapoda, Chilopoda) Zoomorphologie, 84, pp. 77-102.

Keil T.A. and Steinbrecht R.A. (1984) Mechanosensitive and Olfactory Sensilla of Insects. In: King R.C., Akai H. (eds) Insect Ultrastructure, 2, pp. 477-516.

Kladt N., Wolf H., Heinzel H.G. (2007) Mechanoreception by cuticular sensilla on the pectines of the scorpion *Pandinus cavimanus*. Journal of Comparative Physiology A, 193, pp. 1033–1043.

Klagges B.R.E., Heimbeck G., Godenschwege T.A., Hofbauer A., Pflugfelder G.O., Reifegerste R., Reisch D., Schaupp M., Buchner S., Buchner E. (1996) Invertebrate Synapsin: A single gene codes for several isoforms in *Drosophila*. Journal of Neuroscience. 16 (10) 3154-3165.

Krapf D. (1986) Contact Chemoreception of Prey in Hunting Scorpions (Arachnida: Scorpiones. Zoologischer Anzeiger, 217 1/2, pp. 119-129.

Loesel R., Nässel D.R., Strausfeld N.J. (2002) Common design in a unique midline neuropil in the brains of arthropods. Arthropod Structure & Development, 31, pp. 77-91.

Loesel, R., Wolf, H., Kenning, M., Harzsch, S., Sombke, A., (2013) Architectural principles and evolution of the arthropod central nervous system. In: Minelli, A., Boxshall, G., Fusco, G. (ed), Arthropod Biology and Evolution: Molecules, Development, Morphology, pp. 299-342.

McIver S.B. (1975) Structure of cuticular mechanoreceptors of arthropods. Annual Review of Entomology, 20, pp. 381-397.

Melville J.M. (2000) The pectines of scorpions: analysis of structure and function. Dissertation submitted in fulfilment of the requirements for the degree of Doctor of Philosophy at the Oregon State University, USA.

Melville J.M., Tallarovic K., Brownell P.H. (2003) Evidence of Mate Trailing in the Giant Hairy Desert Scorpion, *Hadrurus arizonensis* (Scorpionida, Iuridae). Journal of Insect Behavior, 16, pp. 97–115.

Mineo M.F. and Claro D.K. (2006) Mechanoreceptive function of pectines in the Brazilian yellow scorpion *Tityus serrulatus*: perception of substrate-borne vibrations and prey detection. acta ethologica, 9, pp. 79–85.

Pringle J.W.S (1955) The Function of the Lyriform Organs of Arachnids. Journal of Experimental Biology, 32, pp. 270-278.

Richter S., Loesel R., Purschke G., Schmidt-Rhaesa A., Scholtz G., Stach T., Vogt, L., Wanninger A, Brenneis G, Döring C, Faller S, Fritsch M, Grobe P, Heuer CM, Kaul S, Möller OS, Müller CHG, Rieger V, Rothe BH, Stegner MEJ, Harzsch S (2010) Invertebrate neurophylogeny: suggested terms and definitions for a neuroanatomical glossary. Frontiers in Zoology, 7, pp. 29.

Romer A.S. and Parsons T.S. (1977) The Vertebrate Body. Philadelphia, PA: Holt-Saunders International. pp. 129–145.

Ruppert E.E., Fox R.S., Barnes R.D. (2007) Invertebrate Zoology: A Functional Evolutionary Approach (7th ed.). ISBN:0-03-025982-7. Thomson Learning. pp. 963-980.

Sandeman D.C., Sandeman R.E., Derby C.D., Schmidt M. (1992) Morphology of the Brain of Crayfish, Crabs, and Spiny Lobsters: A Common Nomenclature for Homologous Structures. Biological Bulletin, 183, pp. 304-326.

Schachtner J., Schmidt M., Homberg U. (2005) Organization and evolutionary trends of primary olfactory brain centers in Tetraconata (Crustacea + Hexapoda). Arthropod Structure & Development, 34, pp. 257-299.

Schmidt K. (1969) Der Feinbau der stiftführenden Sinnnesorgane im Pedicellus der Florfliege Chrysopa Leach (Chrysopidae, Plannipennia). Zeitschrift für Zellforschung, 99, pp. 357–388.

Schmidt-Nielsen K. (1984) The strength of bones and skeletons, Scaling: Why is Animal Size So Important? Cambridge University Press, pp. 42–55.

Schmidt M. and Ache B.W. (1992) Antennular projections to the midbrain of the spiny lobster. II. Sensory innervation of the olfactory lobe. The Journal of Comparative Neurology, 318, pp. 291-303.

Schmitt B.C. and Ache B.W. (1976) Olfaction: Responses of a Decapod Crustacean Are Enhanced by Flicking. Science, 205, pp. 204-206.

Schneider D. (1964) Insect antennae. Annual Review of Entomology, 9, pp. 103-122.

Schneider D. and Steinbrecht R.A. (1968) Checklist of insect olfactory sensilla. Proceedings of the Zoological Society of London, 23, pp. 279–297.

Scholtz G. and Edgecombe G.D. (2006) The evolution of arthropod heads: reconciling morphological, developmental and palaeontological evidence. Development, Genes & Evolution, 216, pp. 395–415.

Sharma P.P., Fernández R., Esposito L.A., Gonzalez-Santillan E., Monod L. (2015) Phylogenomic resolution of scorpions reveals multilevel discordance with morphological phylogenetic signal. Proceedings of the Royal Society Biological Science. 282: 20142953. DOI: https://doi.org/10.1098/rspb.2014.2953.

Slifer E.H. (1970) The Structure of Arthropod Chemoreceptors. Annual Review of Entomology, 15, pp. 121-142.

Sombke A., Harzsch S., Hansson B.S. (2011) Organization of deutocerebral neuropils and olfactory behavior in the centipede Scutigera coleoptrata (Linnaeus, 1758) (Myriapoda: Chilopoda). Chemical Senses, 36, pp. 43–61.

Sombke A., Rosenberg J., Hilken G., Westermann M., Ernst A. (2011) The Source of Chilopod Sensory Information: External Structure and Distribution of Antennal Sensilla in Scutigera coleoptrata (Chilopoda, Scutigeromorpha). Journal of Morphology, 272, pp. 1376–1387.

Sombke A., Rosenberg J., Hilken G (2011) Chilopoda - The Nervous System Treatise on Zoology-Anatomy, Taxonomy, Biology. The Myriapoda, Vol. 1, pp. 217-234.

Sombke A., Lipke E., Kenning M., Müller K.H.G, Hansson B.S., Harzsch S. (2012) Comparative analysis of deutocerebral neuropils in Chilopoda (Myriapoda): implications for the evolution of the arthropod olfactory system and support for the Mandibulata concept. BioMed Central Neuroscience,13:1. https://doi.org/10.1186/1471-2202-13-1

Steinbrecht R.A. (1987) Functional morphology of pheromone-sensitive sensilla. Pheromone Biochemistry, pp. 353-384.

Stocker R.F., Lienhard M.C., Borst A., Fischbach K.F. (1990) Neuronal architecture of the antennal lobe in Drosophila melanogaster. Cell and Tissue Research, 262, pp. 9-34.

Stocker R.F. (1994) The organization of the chemosensory system in Drosophila melanogaster: a review. Cell and Tissue Research, 275, pp. 3-26.

Strausfeld N.J., Hansen L, Li Y., Gomez R.S., Ito K. (1998) Evolution, Discovery, and Interpretations of Arthropod Mushroom Bodies. Learn and Memory, 5, pp. 11-37.

Strausfeld N.J. and Hildebrand J.G. (1999) Olfactory systems: common design, uncommon origins? Current Opinion in Neurobiology, 9, pp. 634-639.

Strausfeld N.J. (2009) Brain organization and the origin of insects: an assessment. Proceedings of the Royal Society Biological Science. 276, 1929-1937. DOI: https://doi.org/10.1098/rspb.2008.1471.

Strausfeld N.J., Sinakevitch I., Brown S.M., Farris S.M. (2009) Ground plan of the insect mushroom body: functional and evolutionary implications. Journal of Comparative Neurology, 513, pp. 265-291.

Tallarovic S.K., Melville J.M., Brownell P.H. (2000) Courtship and Mating in the Giant Hairy Desert Scorpion, *Hadrurus arizonensis* (Scorpionida, Iuridae). Journal of Insect Behavior, 13, pp. 827–838.

Tichy H. and Barth F.G. (1992) Fine structure of olfactory sensilla in myriapods and arachnids. Microscopy Research & Technique, 22, pp. 372–391.

Vosshall L.B. and Stocker R.F. (2007) Molecular architecture of smell and taste in Drosophila. Annual Review of Neuroscience, 30, pp. 505-533.

Wachowiak M. and Shipley M.T. (2006) Coding and synaptic processing of sensory information in the glomerular layer of the olfactory bulb. Seminars in Cell & Developmental Biology, 17, pp. 411–23.

Wainwright S.A., Biggs W.D., Gosline J.M. (1982) Mechanical Design in Organisms, Princeton University Press, pp. 162–163.

Weygoldt P. (2000) Whip Spiders (Chelicerata: Amblypygi): Their Biology, Morphology and Systematics. Apollo Books, Stenstrup, Denmark. pp. 1-163.

Wolf H. and Harzsch S. (2002a) The neuromuscular system in the walking legs of a scorpion. 1. Arrangement of muscles and excitatory innervation. Arthropod Structure & Development, 31, pp. 185-202.

Wolf H. and Harzsch S. (2002b) The neuromuscular system in the walking legs of a scorpion. 2. Inhibitory motoneurons. Arthropod Structure & Development, 31, pp. 203-215.

Wolf H. (2008) The pectine organs of the scorpion, *Vaejovis spinigerus*: Structure and (glomerular) central projections. Arthropod Structure & Development, 37, pp. 67-80.

Wolf H. and Harzsch S. (2012) Serotonin-immunoreactive neurons in scorpion pectine neuropils: similarities to insect and crustacean primary olfactory centres? Zoology, 155, pp. 151-159.

Wolf H. (2017) Scorpions pectines - Idiosyncratic chemo- and mechanosensory organs. Arthropod Structure & Development. Arthropod Structure & Development, 46, pp. 753-764.

Wolff G., Harzsch S., Hansson B.S., Brown S., Strausfeld N.J. (2012) Neuronal organization of the hemiellipsoid body of the land hermit crab, *Coenobita clypeatus*: correspondence with the mushroom body ground pattern. Journal of Comparative Neurology, 520, pp. 2824-2846.

Zacharuk R.Y. (1980) Ultrastructure and Function of Insect Chemosensilla. Annual Review of Entomology, 25, pp. 27-47.

Appendix

A 1 Chemicals

The following table comprises all chemicals used and their respective brands.

Chemical	Brand
NaCl	Merck KGaA, Germany
NaH$_2$PO$_4$	Merck KGaA, Germany
Na$_2$HPO$_4$	Merck KGaA, Germany
PFA 95 % Powder	Sigma-Aldrich Co. LLC., Hamburg, Germany
PDL (Poly-d-Lysine) 1mg/ml in H$_2$O	Merck KGaA, Germany
Agarose Low Melt	Carl Roth GmbH + Co. KG, Germany
Saponin	Fluka BioChemika, Honeywell International Inc., USA
Triton™ X-100	Sigma-Aldrich Co. LLC., Hamburg, Germany
NGS (Normal goat serum)	Vector Laboratories, Canada
Neurobiotin™ Tracer	Vector Laboratories, Canada
Alexa Fluor™ 488 goat anti-mouse IgG	Invitrogen by Thermo Fisher Scientific Life Technologies GmbH, Darmstad, Germany
Cy3™-Streptavidin	Invitrogen by Thermo Fisher Scientific Life Technologies GmbH, Darmstad, Germany
DAPI	Sigma-Aldrich Co. LLC., Hamburg, Germany
Mowiol®4-88	Sigma-Aldrich Co. LLC., Hamburg, Germany
Glycerol	Merck KGaA, Germany
Ethanol 99,6 %	Merck KGaA, Germany
DMSO (Dimethyl Sulfoxide)	Sigma-Aldrich Co. LLC., Hamburg, Germany

© Springer Fachmedien Wiesbaden GmbH, part of Springer Nature 2019
D. Drozd, *Topographic Organization of the Pectine Neuropils in Scorpions*,
BestMasters, https://doi.org/10.1007/978-3-658-25155-0

A 2 Solutions

Phosphate Buffered Saline (PBS) 1 l Solution

Chemical	M [g/mol]	n [mol/l]	m [g/l]
NaCl	58.440	145.300	8.491
NaH_2PO_4	137.990	1.500	0.207
Na_2HPO_4	177.990	8.400	1.495

Paraformaldehyde (PFA) 4 %
Recipe for 100 ml:
- Dissolve 4 g of PFA in H_2O bidest. and heat up the solution to 60 °C (under the fume cupboard!).
- Add 1-2 drops of NaOH until solution is clear.
- Add 50 ml of 2x PBS until pH 7.4 is reached.
- Fill up with H_2O bidest. to 100 ml.

A 3 Tools and equipment

The animals (*Mesobuthus eupeus*) used in this study were obtained from the online shop "the pet factory". In total, 35 animals were used for the experiments. Before preparation, the animals were held in tanks, with no differentiation between the sexes.

The following table contains all tools and equipment used for this study

Tool/Equipment	Brand
Cork plates	Self made
Vaseline	Vaseline® Original, Germany
Wax	Aspermühle, Germany
Glass chamber	DWK Life Sciences (WHEATON®), Germany
Glass capillary, GC100TF – 10, 1.0 mm O.D. x 0.78mm I.D.	Harvard Apparatus, USA
Micropipette Puller	Sutter Instrument Co., Model P-97, USA
Disposable Glass Pasteur Pipettes 150 mm	VWR, USA
Falcon 15 ml, 17 x 120 mm style	Corning Incorporated, USA
Falcon 50ml, 30 x 115 mm style	Corning Incorporated, USA
12 well cell culture plate	Corning Incorporated, USA

Vibratom Leica VT1000 S	Leica Biosystems Nussloch GmbH, Germany
Microscope slides SuperFrost®Plus	VWR, USA
Microscope cover glasses 24 x 60 mm	VWR, USA
Dissecting tools	Fine science tools GmbH, Germany
Scale	Sartorius, Germany